2nd International Conference on Robotics and Mechantronics (ICRoM 2018)

IOP Conference Series: Materials Science and Engineering
Volume 517

Singapore
9 – 11 November 2018

ISBN: 978-1-5108-8636-0
ISSN: 1757-8981

Printed from e-media with permission by:

Curran Associates, Inc.
57 Morehouse Lane
Red Hook, NY 12571

Some format issues inherent in the e-media version may also appear in this print version.

Printed by Curran Associates, Inc. (2019)

For permission requests, please contact the Institute of Physics at the address below.

Institute of Physics
Dirac House, Temple Back
Bristol BS1 6BE UK

Phone: 44 1 17 929 7481
Fax: 44 1 17 920 0979

techtracking@iop.org

Additional copies of this publication are available from:

Curran Associates, Inc.
57 Morehouse Lane
Red Hook, NY 12571 USA
Phone: 845-758-0400
Fax: 845-758-2633
Email: curran@proceedings.com
Web: www.proceedings.com

Table of contents

Volume 517

2nd International Conference on Robotics and Mechantronics

9–11 November 2018, Singapore

Accepted papers received: 12 February 2019
Published online: 26 April 2019

Preface

Papers

Material Performance Testing and Analysis

Vehicle Engineering and Modern Manufacturing Technology

Mechatronics and Robot Design

IOP Conf. Series: Materials Science and Engineering **517** (2019) 011001 doi:10.1088/1757-899X/517/1/011001

Preface

Nowadays, Automation and Mechatronics Engineering play an increasingly important role in both control and engineering applications. In the real world, the environment is complex and dynamics. As such, the automation systems should learn and adapt accordingly and more efforts should be focused on the methodology of the learning system on one hand. For example, fast adaptation and self-organizing capability are highly desired and research activities on this type of development should be expedited. On the other hand, one should leverage on artificial intelligence and machine learning to enhance their performance. Towards this end, the 2018 the 2nd International Conference on Robotics and Mechantronics (ICRoM 2018) was organized.

The main goal of ICRoM 2018, which will be held from 9 to 11 November 2018, in Singapore, is to address latest original results in advanced intelligent control of automation systems, including both theoretical advances and practical implementations, which are becoming more and more popular in industry and in our daily lives. The ICRoM 2018 will provide a premier interdisciplinary platform for scientists, researchers, industry leaders, engineers and educators throughout the world to present and discuss the most recent innovations, trends, concerns, as well as practical challenges encountered, and streamline solutions in the fields of Automation and Mechatronics Engineering. The meeting will provide an opportunity to highlight recent developments and to identify emerging and future areas of growth in Robotics and Mechantronics.

Towards this end, the Technical Programme Committee has assembled an excellent programme comprising of 2 excellent Keynote Speeches and 3 Plenary Talks from renowned scientists from the world, 2 parallel technical oral sessions comprising of 18 oral presentations. A total of 13 papers were selected from 41 original contributions after a rigorous review process reflecting a high rejection rate of more than 68 %.

On behalf of the Organising Committee, I wish to thank the keynote speakers, invited speakers and authors of selected papers for their outstanding contributions. I would also like to thank members of the organizing committee, anonymous reviewers and volunteers for their great efforts. Without their contribution, dedication and commitment, we would not have achieved so much. We sincerely hope that you will find the ICRoM 2018 beneficial and fruitful for your professional development. We also hope that you will enjoy our hospitality and will have an enjoyable and memorable time in Singapore.

General Chair, ICRoM 2018
Professor Meng Joo Er
Nanyang Technological University, Singapore

IOP Conf. Series: Materials Science and Engineering **517** (2019) 011002 doi:10.1088/1757-899X/517/1/011002

Conference Committee

International Advisory Committee

Prof. Youmin Zhang, Concordia University, Canada

General Chair

Prof. Meng Joo Er, Nanyang Technological University, Singapore

Program Chairs

Prof. Christos Spitas, Nazarbayev University, Kazakhstan

Prof. Andrew Ragai Henry Rigit, Universiti Malaysia Sarawak, Malaysia

Assoc. Prof. Rosli Bin Ahmad, Universiti Tun Hussein Onn Malaysia, Malaysia

Technical Committees

Prof. Joao Manuel R.S. Tavares, Universidade do Porto, Portugal

Prof. Rene V. Mayorga, University of Regina, Canada

Asst. Prof. Gordon R. Pennock, Purdue University, USA

Prof. Tarek M. Sobh, University of Bridgeport, USA

Prof. Jorge Dias, University of Coimbra, Portugal

Prof. Chong Kok-Keong, Universiti Tunku Abdul Rahman, Malaysia

Prof. Peng-Yeng Yin, National Chi Nan University, Taiwan

Prof. Ahmet Sayar, Kocaeli University, Turkey

Prof. Juan M. Corchado, University of Salamanca, Spain

Prof. Guohua Cui, Hebei University of Engineering, China

Prof. Essam E. Khalil, Cairo University, Egypt

Prof. Garje Channabasappa Mohan kumar, National Institute of Technology Karnataka, India

Asst. Prof. Dr. V. ANANDAKRISHNAN, National Institute of Technology, India

Dr. N. MOHANA SUNDARA RAJU, Mahendra Institute of Technology, India

Dr. Muhamad Mat Noor, University Malaysia Pahang, Malaysia

Dr. Hany Nasry, Military Technical College, Egypt

IOP Conf. Series: Materials Science and Engineering **517** (2019) 011003 doi:10.1088/1757-899X/517/1/011003

Peer review statement

All papers published in this volume of *IOP Conference Series: Materials Science and Engineering* have been peer reviewed through processes administered by the proceedings Editors. Reviews were conducted by expert referees to the professional and scientific standards expected of a proceedings journal published by IOP Publishing.

IOP Conf. Series: Materials Science and Engineering **517** (2019) 012001 doi:10.1088/1757-899X/517/1/012001

Effect of ZrO_2 Addition on Mechanical Properties and Microstructure of Al-9Zn-6Mg-3Si Matrix Composites Manufactured by Squeeze Casting

Muhammad Syarifudin, Elisabeth Nadya Hale, Bondan Tiara Sofyan
Department of Metallurgy and Materials Engineering, Faculty of Engineering, Universitas Indonesia, Kampus UI Depok 16424, Indonesia

Abstract. Steel is used because of its high strength and toughness, but it has high density, therefore lighter material with comparable toughness is developed. One alternative is aluminum matrix composite with zirconia (ZrO_2) as the reinforcement with high fracture toughness. Al-9Zn-6Mg-3Si (wt. %) composites were developed with addition of 2.5, 5, and 7.5 vol. % ZrO_2 through squeeze casting. To improve toughness, the composite was solution treated at 450 °C for 1 h, then aged at 200 °C for 1 h. Materials characterization included Optical Emission Spectroscopy (OES), Rockwell B hardness testing, impact testing, fractography analysis, microstructure analysis using Optical microscope (OM) and Scanning Electron Microscope (SEM) / Energy Dispersive X-Ray Spectroscopy (EDS), as well as X-Ray Fluorescence (XRF). The results showed that the more ZrO_2 particles, the higher porosity and the lower the hardness and the impact values, both in as-cast condition and after ageing at 200 °C at 1 h.

1. Introduction

Steel which is known to have high strength and high toughness is commonly used in several applications. Still, the high specific mass of steel (7.85 g/cm^3) makes it less effective. Therefore, metal matrix composites (MMC) with light matrix and high toughness reinforcement are developed as alternative materials.

Aluminium 7xxx contains zinc, magnesium, and copper as its main alloying elements. It is suitable to be used as composite matrix considering its low density along with superior strength and toughness compared to the other aluminium alloy series. Zirconia (ZrO_2) is used as the reinforcement to promote the fracture toughness and strength due to its value of fracture toughness of 13 MPa.m$^{-0.5}$, Young Modulus of 190 GPa, and hardness of 1250 HV [1].

Zirconia may increase the strength of the aluminium matrix composite with two mechanisms, the first one being grain refinement and the second one being the increment of dislocation density as a result of the difference in thermal expansion coefficient between ZrO_2 and aluminium. Previous research conducted by Madhusudhan [2] showed that the addition of up to 8% ZrO_2 particles in Al-8Zn-3Mg-2Cu alloy increased the hardness and strength. However Kharizma [3] explained that higher addition of ZrO_2 on Al-13.1Zn-6.1Mg-6.7Si-1.4Cu alloy, resulted in higher porosities due to the low wettability of ZrO_2 particles in molten aluminium [4].

The presence of porosity will lower the strength and hardness of the aluminium matrix composite, therefore squeeze casting should be used to enhance heat transfer through the contact between the molten metal with the pressurized hydraulic plate. It will result in fine grained products with minimum porosity and shrinkage. Yang [5] concluded that squeeze casting will reduce the solidification time of the molten aluminium that improved mechanical properties. Another research by Shi et. al [6] proved that the optimum pouring temperature for squeeze casting method of aluminium alloy is 690~700 °C that produced the highest tensile strength and elongation, since the grains are not overly grown. This research studies Al-9Zn-6Mg-3Si composites with addition of 2.5, 5, and 7.5 vol. % ZrO_2 particles produced by squeeze casting. Further ageing at 200 °C for 1 h aimed to increase toughness and hardness without significantly embrittling the materials. Since ageing can generate second phases which are intermetallic compounds, specifically $MgZn_2$ in Aluminium 7xxx. These intermetallic compounds may appear in different shapes including needle-like, oval, or even Chinese-script which increase the toughness of the aluminium composites.

2. Materials and Method

This study used Al-9Zn-6Mg-3Si (wt. %) alloy as the matrix and ZrO_2 as the reinforcement, with the variation of 2.5, 5, and 7.5 (vol. %). Input materials were Al, Al-20Si, Zn and Mg ingots that were melted in a crucible furnace at 850 °C. Degassing was conducted by using argon gas for 60 seconds. Zirconia particles were previously heated at 1100 °C and then added to the molten metal while stirred for 30 s. The molten composite was poured in 170x70x55 mm^3 metal mould and then squeezed at 76 MPa for 10 min. The composites were then homogenized at 450 °C for 24 h. Table 1 presents the obtained composition of the aluminium composites tested by Optical Emission Spectroscopy (OES) Thermo Scientific ARL 2000.

Rockwell hardness test were conducted in accordance to ASTM E18 with three indentation for each measurement. Charpy impact test referred to ASTM E23 standard and metallographic standard preparation was used with 0.5 % HF etchant. Microstructures were obtained by using Optical Microscope (OM), SEM-EDS and analyzed by ImageJ Software. X-Ray Fluorescence (XRF) Portable X-Spectra, and X-Ray Diffraction (XRD) tests were also performed.

The samples were then solution treated at 450 °C for 1 h followed by water quenching and aged at 200 °C for 1 h. The aged samples were also characterized with the same methods.

Table 1. The actual composition of the aluminium composite

Element	Composition (wt. %)			
	Composite A (0 % ZrO_2)	**Composite B (2.5 % ZrO_2)**	**Composite C (5 % ZrO_2)**	**Composite D (7.5 % ZrO_2)**
Zn	9.66-9.72	9.98-10.36	9.69-9.78	10.17-10.36
Mg	5.27-5.39	9.66-10.51	4.55-4.61	4.66-5.07
Si	2.65-2.67	2.49-2.51	2.64-2.66	2.57-2.58
Al	*Bal.*	*Bal.*	*Bal.*	*Bal.*

3. Results and Discussion

3.1 Microstructural Characterization

Figs. 1 and 2 show that with higher ZrO_2 particles content, the more porosities found in the microstructures in both as-cast and aged Al-9Zn-6Mg-3Si composites, which are indicated by arrows in the microstructure. In Fig. 2 the yellow arrows show the silicon interdendritic phases and the green arrows point to chinese script structures which are the intermetallics. Meanwhile the red and purple arrows are shrinkages and porosities, respectively, that increase by the addition of ZrO_2 particles. The porosities appear to be much larger as the ZrO_2 content higher.

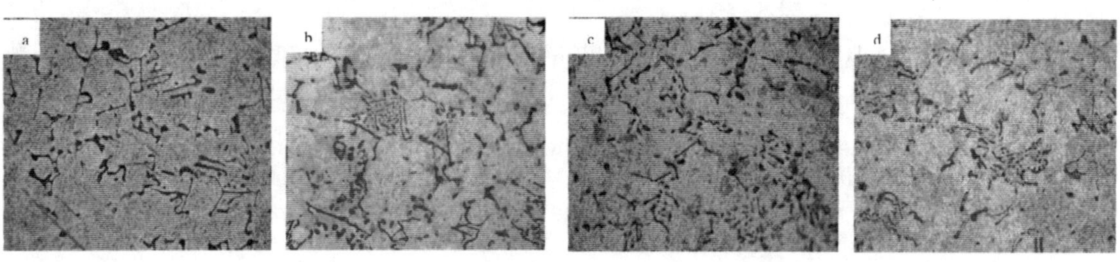

Figure 1. Microstructure of as-cast Al-9Zn-6Mg-3Si with (a) 0, (b) 2.5, (c) 5, and (d) 7.5% ZrO_2 particles.

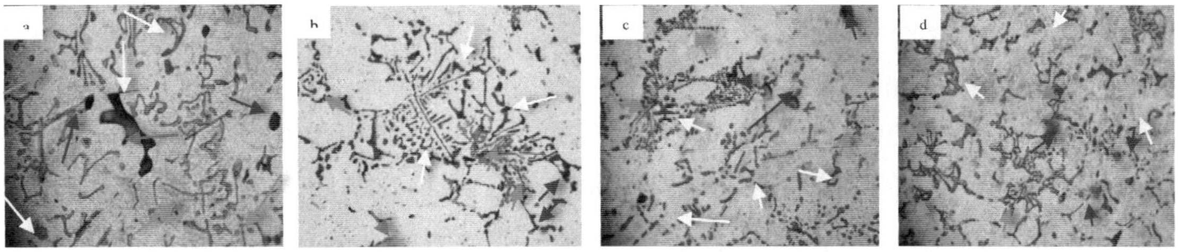

Figure 2. Microstructure of aged Al-9Zn-6Mg-3Si with (a) 0, (b) 2.5, (c) 5, and (d) 7.5% ZrO_2 particles. Yellow, green, red and purple arrows indicate silicon, chinese script, shrinkages and porosities, respectively.

SEM micrographs of Al-9Zn-6Mg-3Si composites with 0 and 2.5 % ZrO_2 particles are displayed in Fig. 3 while Table 2 presents EDS characterization at complementary positions. In both figures, EDS examination did not show any content of ZrO_2 particles. However, in Table 3, which presents XRF characterization in the composite with 2.5 vol. % ZrO_2, Zr element is detected, proving the ZrO_2 reinforcement is there. This difference take place as a result of uneven ZrO_2 distribution, for some parts of ZrO_2 particles gathered in a porosity and the others is settled down inside the crucible.

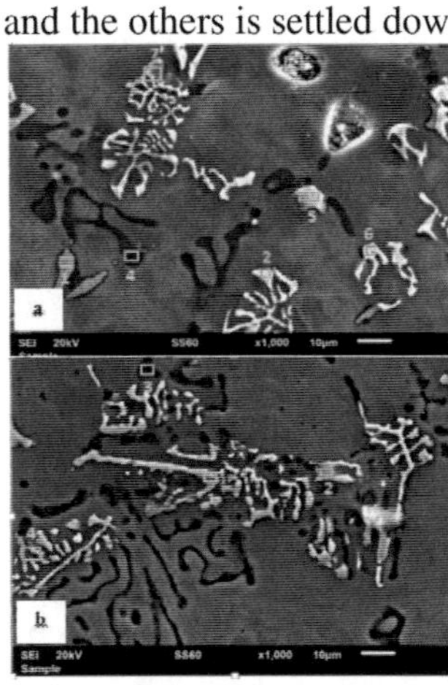

Figure 3. SEM micrographs of the aged Al-9Zn-6Mg-3Si composite with (a) 0 and (b) 2.5 vol % ZrO2 particles

Table 2. Microanalysis at points shown in Fig. 3(a)

Point	Element (% at.)							Phase
	Zn	Mg	Si	O	C	Fe	Al	
1	10.89	-	-	-	-	-	89.11	Al
2	7.41	14.66	10.92	8.62	-	0.18	58.21	$FeSiAl_8$
3	10.43	-	-	3.31	-	-	86.26	Al_2O_3
4	10.09	2.43	-	4.72	-	-	84.53	$MgZn_2$
5	8.51	6.64	3.88	4.72	-	-	76.25	Mg_2Si
6	5.01	2.02	5.04	4.7	15.66	16.65	50.01	Mg_2Si
7	6.45	-	1.98	6.32	23.04	8.12	54.13	$FeAl_3$

Table 3. Microanalysis at points shown in Fig. 3(b)

Point	Element (% at.)							Phase
	Zn	Mg	Si	O	C	Fe	Al	
1	4.72	3.92	4.94	6.15	12.94	9.65	57.49	$Mg_3Zn_3Al_2$
2	4.41	10.21	5.54	5.64	11.1	8.36	54.75	$MgZn_2$
3	1.36	46.49	25.47	5.43	-	-	21.25	Mg_2Si
4	6.6	-	1.69	5.05	16.49	6.28	63.99	$FeAl_3$

The chinese script morphology is proven to be the intermetallic Mg_2Si and $FeSiAl_8$ phases (points 2, 5, 6 in Fig. 3(a) and point 3 in Fig. 3(b)). Other intermetallics were $Mg_3Zn_3Al_2$ (point 1 in Fig. 3(b)) and $MgZn_2$ (point 4 in Fig. 3(a) and point 2 in Fig. 3(b)). According to Hirsch, et. al [7], these intermetallic phases have a great effect in the formation of porosity since they will reduce metal fluidity and enhance the chance of micro porosity formation.

3.2 Mechanical Properties

The hardness of the Al-9Zn-6Mg-3Si composites with 0, 2.5, 5, and 7.5 % ZrO_2 reinforcement is shown in Fig. 4. Overall, Fig. 4 indicates a decrease in the hardness along with the increase in ZrO_2 content, both in the as-cast and aged condition. Samples in aged condition possessed higher hardness values, this happens because after the ageing treatment, precipitation hardening mechanism occurred.

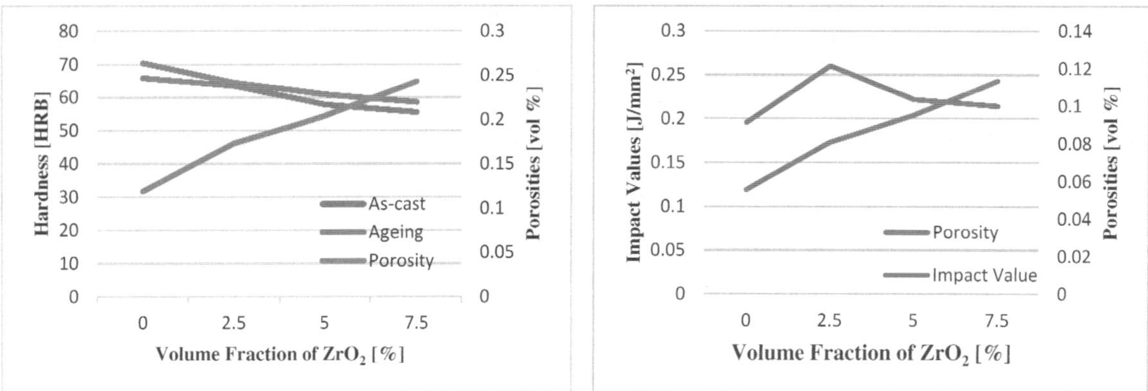

Figure 4. Effects of ZrO_2 on, a) the hardness and b) impact values of Al-9Zn-6Mg-3Si composites in as-cast condition and after ageing at 200 °C for 1 h.

These results are affected greatly by the presence of the porosity in the matrix. To confirm the amount of porosities in the samples quantitative analysis was conducted and the results are presented in Fig.4. The porosities are larger and in higher amount with the addition of ZrO_2 particles. These results are in line with previous work [8], which found that addition of ZrO_2 for more than 5% in Al-9Zn-6Mg-3Si composites diminished the strengthening effects due to the presence of porosities. With the increasing porosity, it is confirmed in Fig. 4 that the hardness value of the composites which contains more ZrO_2 particles is lower than the composites with the less ZrO_2 particles. This explains the great difference between the actual hardness and theoretical hardness calculated with the rules of mixture shown in Table. 4 that indicates the effect of the porosities is very dominant compared to the strengthening effect of reinforcement and ageing.

Table 4. Calculated results of theoretical and actual as-cast and aged composites hardness

Volume Fraction of ZrO_2 [*vol.%*]	Actual Hardness [HRB]	Theoretical Hardness without porosity [HRB]
0	65.85	65.93
2.5	63.45	92.98
5	57.82	120.03
7.5	55.47	147.08
Volume Fraction of ZrO_2 [*vol.%*]	**Actual Hardness [HRB]**	**Theoretical Hardness without Porosity [HRB]**
0	70.37	70.56
2.5	64.45	97.50
5	60.87	124.43
7.5	58.53	151.37

The result shown in Fig. 4 displays an overall decrease in the impact value. This supports the research conducted by Aqida, et. al [9] that the presence of porosity will cause a reduction in the toughness of the materials. This happens because the porosities are the crack initiation and propagation path when the load is given so the energy absorbed is lower.

The fractography images of the Al-9Zn-6Mg-3Si composites in Fig. 5 show brittle fractures in all 0, 2.5, 5, and 7.5 vol% ZrO_2.

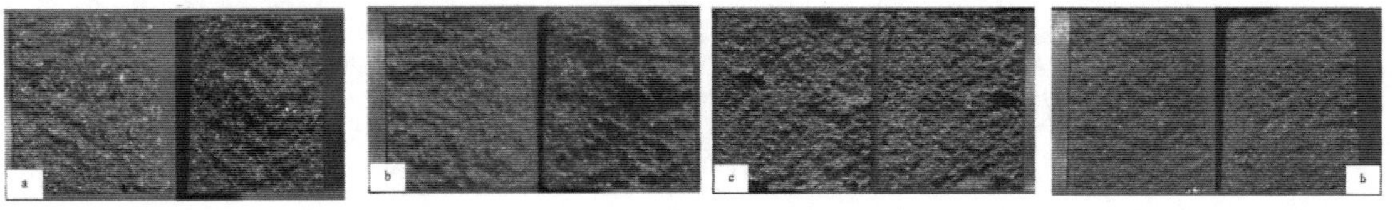

Figure 5. The macro fractography impact testing result of Al-9Zn-6Mg-3Si composites with (a) 0, (b) 2.5, (c) 5, and (d) 7.5 vol% ZrO_2 particles.

4. Conclusions
The study of ZrO_2 reinforcement addition with variation of 0, 2.5, 5, and 7.5 vol% in Al-9Zn-6Mg-3Si composite resulted in the following conclusions :

1) Addition of ZrO_2 particles correlated with formation of porosities so that reduced the hardness and impact values of the composites.
2) Ageing at 200 ^{0}C for 1 h increased the hardness of the composites which may be due to the formation of precipitates.
3) The ZrO_2 particles were not distributed evenly because of the poor wettability with the molten Aluminium resulting in the formation of porosity in the composite.
4) Special treatment for the ZrO_2 particles is needed to reduce porosity including addition of wetting agent.

5. Acknowledgement

The authors are grateful with the provision of research funding through *Hibah Publikasi Internasional Terindeks untuk Tugas Akhir* (PITTA) 2018 from Universitas Indonesia with contract no. 2401/UN2.R3.1/HKP.05.00/2018.

6. References
[1] J. Park, Bioceramics: Properties, Characterizations, and Applications (2008) USA: Springer Science-Business Media, LLC.
[2] M. Madhusudhan, G. J. Naveen, K. Mahesha, Mechanical Characterization of AA7068-ZrO_2 reinforced Metal Matrix Composites . Mat. Today Proceedings. 4 (2017) 3122-3130.
[3] R.K. Agrista, Studi Pengaruh Penambahan Penguat ZrO_2 Terhadap Karakteristik Komposit Al-13.1Zn-6.1Mg-6.7Si-1.4Cu Hasil Squeeze Casting Untuk Aplikasi Balistik (Study of ZrO_2 Reinforcement Particles Addition on Characteristics of Al-13.1Zn-6.1Mg-6.7Si-1.4Cu Composites Manufactured by Squeeze Casting for Ballistics Application). (2014). Skripsi. Depok: Universitas Indonesia.
[4] S.A. Khorramie, M. A. Baghchesara, D. P. Gohari, Fabrication of aluminum matrix composites reinforced with Al_2ZrO_5 nano particulates synthesized by sol–gel auto-combustion method. Trans. Nonfer. Met. Soc. China. 23 (2013) 1556-1562.
[5] L.J. Yang, The effect of solidification time in squeeze casting of aluminium and zinc alloys. Mat. Proc. Tech. 192-193 (2007) 114-120.
[6] W. Shi, , Squeeze Casting Process and Mould Design for 2A12 Aluminum Alloy Shell. Spec. Cast. And Nonfer. Alloys. 38 (2018) 629-631.
[7] J. Hirsch, B. Skrotzki, G.Gottstein, Aluminium Alloys, Their Physical and Mechanical Properties (2002) New Jersey : John Wiley & Sons, Inc.

[8] H. Abdizadeh, M.A. Baghchesara, Investigation on mechanical properties and fracture behavior of A356 aluminum alloy based ZrO_2 particle reinforced metal-matrix composites. Cer. Int. 39 (2013) 2045-2050.

[9] S.N. Aqida, M.I. Ghazali, J. Hashim, Effects of porosity on mechanical properties of metal matrix composite: an overview. Jur. Tek.. 40 (2004) 17-32.

IOP Conf. Series: Materials Science and Engineering **517** (2019) 012002 doi:10.1088/1757-899X/517/1/012002

Effect of Cold Rolling and Annealing Temperature on the Characteristics of Cu-28Zn-3.2Mn Alloy

William Horizon, Syarah Khayrun Nisa and Bondan Tiara Sofyan

Department of Metallurgy and Materials Engineering, Faculty of Engineering, Universitas Indonesia, Kampus UI Depok 16424, Indonesia

Email: bondan@eng.ui.ac.id

Abstract. Common material for bullet shell is cartridge brass which contains 26-32 wt. % Zn. In the deep drawing process, problems are typically found, such as cracking and tearing, due to low ductility. Therefore, manganese is added to cartridge brass to increase its ductility. In this study, Cu-28Zn alloy with addition of 3.2 wt. % Mn was fabricated by gravity casting. As cast samples were homogenized at 800 °C for 2 hours. Afterwards, the specimens were cold-rolled with deformation of 20, 40, and 70 %. The 70 % cold-rolled samples were subsequently annealed at 350, 400, and 450 ℃ for 15 minutes. Samples underwent characterizations by microstructure analysis using optical microscope and Scanning Electron Microscope (SEM) - Energy Dispersive Spectroscopy (EDS), and Vickers microhardness testing. The results showed that higher degree of deformation led to more elongated grains with increasing values of L/D ratio, and higher hardness. Annealing at 350 ℃ for 15 min did not change the deformed microstructures which indicated the stage of recovery and stress relieve. Meanwhile, higher annealing temperatures of 400 and 450 ℃ led to recrystallization and grain growth, respectively. Hardness declined with the increase in annealing temperature. Mn increases the hardness and recrystallization temperature.

1. Introduction

Cartridge shell for small caliber ammunition is commonly fabricated using cartridge brass that contains 26-32 wt. % Zn, due to its excellent properties such as yield strength, tensile strength, ductility, hardness. Cartridge brass is produced by a series of processes, such as casting, rolling, and deep drawing, wherein deep drawing process, cracking and tearing are typically found [1]. To minimize these problems, addition of manganese was found to be essential because it enhances ductility. Increased strength and ductility are necessary to produce good and flawless cartridge shell [2].

In a study conducted by Ovat, et al.[2], the addition of Mn to Cu-28Zn alloy improved the mechanical properties and the crystallographic texture by the grain refining mechanism. The addition of Mn up to 5 % increased formability at high temperatures. However, if the Mn was added more than 5 %, there will be a phenomenon of embrittlement. In addition, studies conducted by Sofyan and Basori [3] on the microstructure and mechanical properties of the annealing and process of the Cu-32Zn alloys were carried out with deformation variations of 20, 40, and 70 % resulted the change of deformation mechanism as increasing deformation degree. The deformation mechanism started with slip at the level of deformation of 20 %, followed by twinning at 40 % deformation, and shear band started to be dominant at 70 % deformation. The mechanisms are affected by Stacking Fault Energy (SFE). Materials with high SFE tend to have homogenous deformation by slip mechanism at major orientation. Whereas,

materials with lower SFE tend to take twinning mechanism that lead inhomogeneous deformation by shear band to form [3].
In the study of Mazancova and Mazanec [4] about SFE of high manganese alloys indicated that carbon and iron do not significantly affect the SFE, in contrast to manganese and aluminum content, where with an increase in Mn, SFE values also increase, which will affect the deformation mechanism of the alloys. The annealing process after cold rolling will yield properties and microstructure as required. Moreover, formability and ductility that declined during cold rolling may be recovered during annealing process [3]. Therefore, this study was carried out to investigate the characteristics of Cu-28Zn-3.2Mn alloys that are subjected to cold-rolling with deformation of 20, 40, and 70 %. Furthermore, the effects of annealing process with temperatures of 350, 400, and 450 °C were also observed.

2. Experimental Method

The samples used in this research are Cu-28Zn alloy with addition of 3.2 wt.% Mn fabricated by gravity casting at melting temperature of 1150 °C, by using pure Cu and Zn ingots, as well as Mn flakes. The molten metal was poured into a 800 °C preheated metal mold with the dimension of 110x110x6 mm^3. In this casting process, borax flux was used to remove scale. Degassing process is not required since Zn would boil away and would bring out trapped gas. The as-cast alloys were then homogenized at 800 °C for 2 hours, and followed by air cooling. The section of samples were then cold rolled at level of deformation of 20, 40, and 70 % in multiple passes. The samples with 70 % deformation were then annealed at 350, 400, and 450 °C for 15 minutes, followed by water quenching.

Microstructures were observed by using optical microscope and SEM/EDS. Standard metallographic preparation was conducted by polishing with 0.5 μm Al_2O_3 and etching with 10 % of $FeCl_3$ + alcohol for 3-6 seconds. Microvickers hardness was performed using 300 g of load in accordance with ASTM E384. Five indentation was made for each measurement.

3. Results and Discussion

3.1. Effects of Deformation

Fig. 1 shows the microstructures of Cu-28Zn-3.2Mn alloy in as-homogenized condition and after a series of deformation. The corresponding hardness values are showed in Fig. 2(a). The as-homogenized microstructure possesses equiaxed large grains with L/d (length to diameter) ratio of ~ 0.7 (Fig. 1a). Some gas porosities were observed due to hydrogen contamination. This microstructure correlates to the hardness value of 56 HV (Fig. 2), which is higher than Cu-32Zn base alloy, 54 HV [3]. The increased hardness is due to the presence of Mn that fully dissolves in α matrix and creates solid solution strengthening [5]. Deformation for 20 % (Fig. 1 (c-d)) significantly increases the grain elongation to L/d ratio to 2.2. Some slip bands are clearly visible, as portrayed by arrows in Fig. 1(d). The hardness of the Mn-added alloy at this stage is 127 HV (Fig. 2), which significantly higher than that of the Cu-32Zn (92 HV) base alloy [3]. This indicates solid solution strengthening mechanism works together with strain hardening.
At 40 % deformation (Fig. 1 (e-f)), the grain elongation increased to L/d 7.7. In FCC crystal, the dislocations move on {111} plane, meaning that they can switch from a plane to another within the same family. It is called cross slips which are shown by arrow in Fig. 1(f) as the dominant deformation. It occurred when perpendicular slip cross one to another. The hardness at this stage is 145 HV (Fig. 2). Further deformation at 70 % (Fig. 1 (g-h)) produced shear bands as shown in Fig. 1(h) with hardness value of 207 HV (Fig. 2). This hardness value still remains higher than that of the Cu-32Zn (192 HV) [3].
Furthermore, Fig. 1(b) shows the microstructure of the as-homogenized sample consists of insoluble dispersoids within the α matrix. However, the EDS micro analysis (Table 1) shows that the particles are Al_2O_3 coming from polishing substances that were not perfectly cleaned. Points 2-3 on small white particles scattered on α matrix are found not much different from the surrounding matrix, the α brass phase.

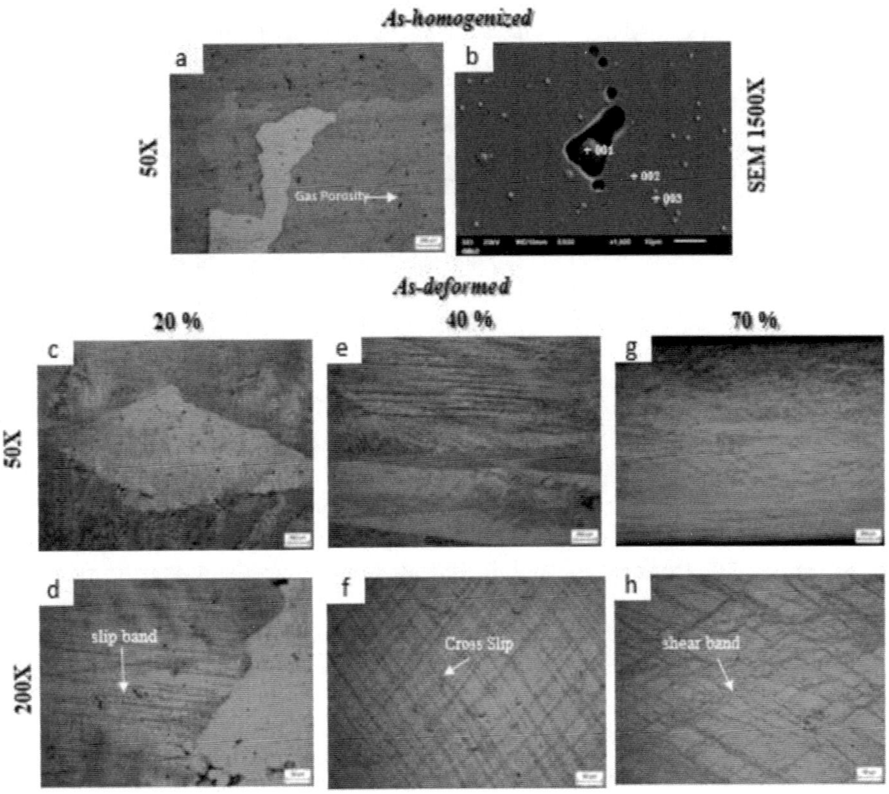

Figure 1. Micrographs of Cu-28Zn-3.2Mn in (a-b) as-homogenized condition, and after deformation for (c-d) 20 % (e-f) 30 % and (g-h) 70 %.

Table 1. Results of EDS microanalysis on Cu-28Zn-3.2Mn alloy before cold rolled at positions as shown in Fig. 1(b)

Point	Composition (%)					Phase
	Cu	Zn	Mn	Al	O	
001	0.52	-	-	40.53	58.95	Al_2O_3
002	68.34	27.76	3.89	-	-	α brass
003	68.73	27.54	3.73	-	-	α brass

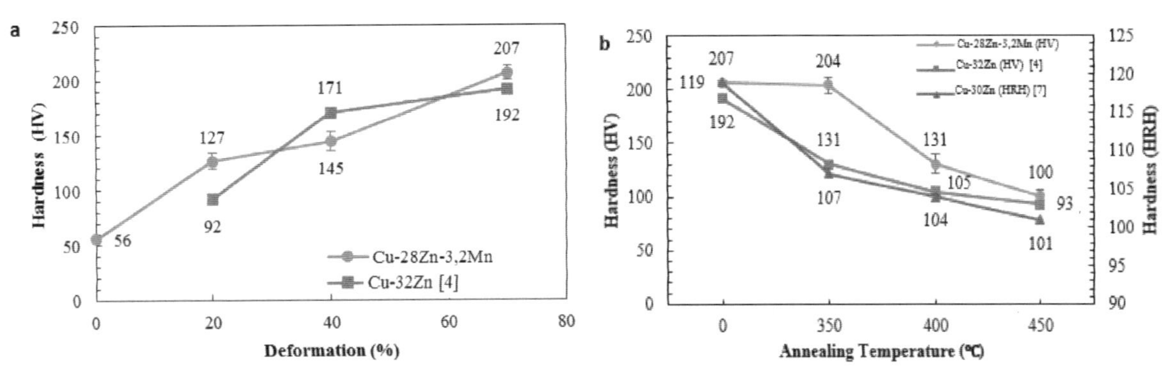

Figure 2. after annealed at 350, 400, 450 °C for 15 min.

3.2. Effects of Annealing Temperature

Fig. 3 illustrates the microstructures after annealing at different temperatures. The corresponding hardness values are showed in Fig. 2(b). Annealing process reduced the hardness of the material. At 350 °C, there was no new nucleation of grains and the existence of the shear band remains visible with a slight decline in hardness to 204 HV (Fig. 2(b)). This reflects the occurrence of recovery where some of the stored strain energy is released by climbing of dislocations [6]. At an increased annealing temperature of 400 °C, the hardness drops to 131 HV (Fig. 2(b)). It can be ascertained that is some parts of the shear band region turned into partial recrystallization as shown in Fig. 3(e-f). The growth of new strain-free granules and equiaxed grain with driving force derived from the internal energy due to deformation [6]. At this stage, the state of the material is similar with that of prior to deformation. Lastly, increasing annealing temperature to 450 °C resulting in a decrease in hardness to 100 HV (Fig. 2(b)). The microstructures (Fig. 3(g-h)) show that the entire shear band region has been replaced by new grains through the recrystallization phenomenon. Some grain growth was observed, in which larger grains tend to make small grains shrunk and seem to be ingested by larger grains (cannibalism) [6]. Larger grain size means less grain boundaries, thus providing more room for the movement of dislocations that leads to decreased hardness and increased ductility. The annealing temperature to achieve complete recrystallization is predicted to be in the range of 400-450 °C.
By comparing the results of this study to previous research by Sofyan and Basori [3] as well as Walker [7] on the annealing treatment of Cu-32Zn, it is noted that addition of 3.2 wt.% increased the recrystallization temperature from 350 °C to 400 °C. Mn atoms induce lattice distortion that may delay movement of dislocation to the grain boundary so that slowing down the recrystallization phenomenon. Therefore higher annealing temperature is needed to achieve perfect recrystallization.

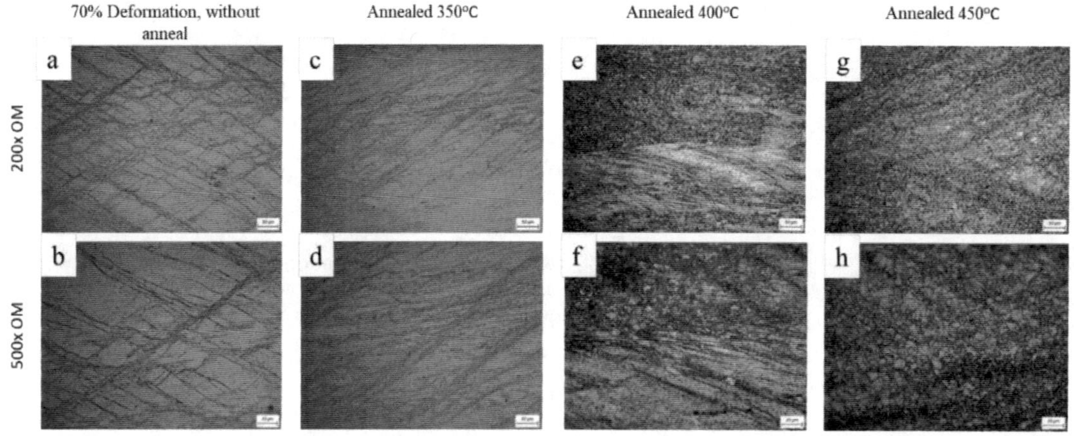

Figure 3. Micrographs of Cu-28Zn-3.2Mn after (a-b) 70 % deformation, and after subsequent annealing at (c-d) 350 °C, (e-f) 400 °C, and (g-h) 450 °C for 15 min.

4. Conclusions
The results of the observation on Cu-28Zn-3.2Mn alloys can be concluded as follows:
1. Addition of 3.2 wt.% Mn resulted in a α single phase with higher hardness due to solid solution strengthening.
2. Mechanism works at 20 % deformation was slip, while higher deformation at 40 % produced cross-slip. Further deformation of 70 % was dominated by shear bands.
3. Increased level of deformation of 20, 40 and 70 % led to higher elongated grains with L/d ratios of ~2.2, ~7.7, and ~14.1 as well as higher hardness values of 127, 145, and 207 HV, respectively.
4. Increased temperatures of the annealing process of 350, 400, and 450 °C decreased the hardness due to the mechanism of recovery and stress relieve, recrystallization, and grain growth, respectively.
5. Addition of Mn increased the recrystallization temperatures, which was found in the range of 400 °C.

IOP Conf. Series: Materials Science and Engineering **517** (2019) 012002 doi:10.1088/1757-899X/517/1/012002

5. Acknowledgement
This research was funded by Penelitian Internasional Terindeks untuk Tugas Akhir (PITTA) 2018 Grant from Universitas Indonesia, with contract No. 2402/UN2.R3.1/HKP.05.00/2018.

6. References

[1] Khanl, N. Z., and K. Azam. *Manufacturing defects of brass products and suggested remedies*, **2(9)**, p. 497–514, 2015.
[2] F. A. Ovat, L. O. Asuquo and F. I. Abam, *The influence of aluminum and manganese on some mechanical properties of brass*, Res. J. Eng. & App. Sci., **1(4)**, p. 214-218, 2012.
[3] B. T. Sofyan and I. Basori, *Effects of deformation and annealing temperature on the microstructures and mechanical properties of Cu-32%Zn Brass*, ARPN J. Eng. App. Sci., **11(4)**, p. 2741-2745, 2016.
[4] E. Mazancova and K. Mazanec, *Stacking fault energy in high manganese alloys*, Materials Eng., **16(2)**, p. 26-31, 2009.
[5] I. Basori, H. I. Pratiwi, and B. T. Sofyan, , *Effects of manganese on the microstructures, mechanical properties and deformation characteristics of Cu-29Zn alloy,* Materials Sci. Forum, **917**, p. 212-217, 2018.
[6] W. D. Callister, Jr. and D. G. Rethwisch, Materials Science and Engineering: An Introduction, New York: John Wiley & Sons, Inc., 1997.
[7] H. L. Walker, *Grain Sized Produced by Recrystallization and Coalescence in Cold-Rolled Cartridge Brass*, Note, P. Urbana: ILL INO, 1945.

IOP Conf. Series: Materials Science and Engineering **517** (2019) 012003 doi:10.1088/1757-899X/517/1/012003

A Study on the Modeling Consideration Dynamic Properties of Vibration Damper Rubber Parts

Tomoaki Kodama[1,a], and Yasuhiro Honda[2,b]

1Department of Science and Engineering, School of Science and Engineering, Kokushikan University, 4-28-1 Setagaya Setagaya-ku Tokyo 154-8515 Japan

2Department of Science and Engineering, School of Science and Engineering, Kokushikan University, 4-28-1 Setagaya Setagaya-ku Tokyo 154-8515 Japan

[a]kodama@kokushikan.ac.jp, [b]honda@kokushikan.ac.jp

Key Words: Damper Pulley; Torsional Vibration; Crankshaft System; Torsional Stiffness; Damping Coefficient; Strain Rate; Loss Factor; Modeling; Rubber; Experiment; Diesel Engine.

Abstract. In this paper, we first describe the dynamic characteristics of damper rubber parts that are necessary for the modeling of torsional vibration. Secondly, we describe an experiment in which two crankshaft pulleys with a torsional rubber damper are fitted to a 6-cylinder, high-speed diesel engine. The torsional waveforms of the damper inertia ring and the pulley are measured by means of phase-shift torsiograph equipment. The measured waveforms are harmonically analyzed and the dynamic characteristics of the stiffness and the damping are investigated from an experimental viewpoint. As a result of comparisons with experimental data, certain dynamic characteristics of pulleys with a torsional rubber damper have been clarified.

1. Introduction

The important requirements in the design stage are that the vibration and noise levels of automobile engines be reduced and that the timbre be improved further [1]-[4]. Torsional vibration dampers of high performance, and crankshaft pulleys with a torsional rubber damper (hereafter called "damper pulleys"), which can reduce both torsional and bending vibrations, and flywheels with a torsional rubber damper have been widely employed in automobile engines [5]-[7]. When designers estimate the amplitude of angular displacement of crankshafts with the abovementioned damper pulleys in the design process, the following issues are of concern:

(1) the estimation of dynamic torsional stiffness and damping of the damper.

These dynamic characteristics are indeterminable [1]-[4]. Therefore, an experimental equation for determining the values has not existed until now and the quantitative values are inaccurate [8]. Unless these values can be accurately estimated to a certain degree, it is difficult to achieve high-precision computation [9]-[13]. We clarify some of the complicated dynamic characteristics of the rubber vibration isolator by measuring the torsional vibrations of the crankshaft system with a damper pulley.

2. Notations

The main symbols used in this paper are as follows:

$\boldsymbol{b}_{RUB}$: Thickness of the rubber at the representative radius [m], $\boldsymbol{C}_{RDP}$: Damping coefficient of the rubber part of a rubber damper pulley [Nms/rad], $\boldsymbol{I}_{RDP}$: Inertia moment of the damper inertia ring [kgm^2], $\boldsymbol{K}_{RDP}$: Dynamic torsional stiffness of the rubber part of a rubber damper pulley [Nm/rad], $\boldsymbol{K}^*_{RDP}$:

Complex torsional stiffness of the rubber part of a rubber damper pulley [Nm/rad] ($= K_{RDP} + j \cdot C_{RDP} \cdot \omega, j$: imaginary number), $\left|K^*_{RDP}\right|$: Absolute torsional stiffness of the rubber part of a rubber damper pulley [Nm/rad] ($= \sqrt{K_{RDP}^2 + (C_{RDP} \cdot \omega_{RDP})^2}$), l_{RDP} : Loss factor ($= C_{RDP} \cdot \omega_{RDP} / K_{RDP}$), M_{RDP} : Amplitude ratio ($= \theta_{RDP,O} / \theta_{PUL,O}$), $r_{RDP,2}, r_{RDP,1}$: Outer and inner diameters of the rubber part, respectively [m], γ_{RDP} : Strain rate [l/s], θ_{RDP} : Torsional angular displacement of the damper inertia ring [rad], θ_{PUL} : Torsional angular displacement of the pulley [rad], θ_{Rel} : Relative torsional angular displacement of the rubber part of a rubber damper pulley [rad] ($= \theta_{RDP} - \theta_{PUL}$), ρ_{RDP} : Representative radius of the rubber part of a rubber damper pulley [m], ϕ_{RDP} : Phase angle between the damper inertia ring and pulley [rad], ω_{RDP} : Angular frequency of torsional vibration [rad/s].

3. Experimental Equipment and Method

This section concerns the main specifications of the test engine and damper pulley, as well as the method of measuring the torsional vibration waveforms. In the experiment, the engines are operated under full load to cause steady forced torsional vibration

3.1 Main Specifications of Test Engine and Rubber Damper Pulley. The test engine is a 6-cylinders, in-line, high-speed diesel engine. Figure 1 shows the dimensions of the test damper pulley. This Table also shows the static stiffness and natural frequency obtained from the static and free vibration tests, respectively. Types A and B of the damper pulleys differ in Shore hardness but their rubber parts have the same shape and dimensions. The materials are similar-quality rubber.

3.2 Method of Measuring Torsional Vibration Waveform. An eddy current dynamometer was connected via a universal joint to the flywheel of the engine. The test damper pulley was fitted to the end of the crankshaft. The gears for generating signal pulses were mounted on the damper inertia ring and pulley. The electric frequency signals proportional to engine speed were obtained from the electromagnetic pickup. The measured signals were transmitted to the phase-shift torsiograph equipment via the adapter which numerical calculated the average of angular velocity (the center frequency). The torsional vibration waveforms could be obtained from the torsional angles, which were numerical calculated using the relationship between the measured and center frequencies. The measured torsional waveforms of the damper inertia ring and pulley were harmonically analyzed using the F.F.T. analyzer. The torsional angular displacement was measured under full load from 800 [r/min] to 3200 [r/min]. The indicator diagrams, data from which were necessary for the vibration analysis described later, were measured using the piezotype indicator in the sixth cylinder from the pulley side.

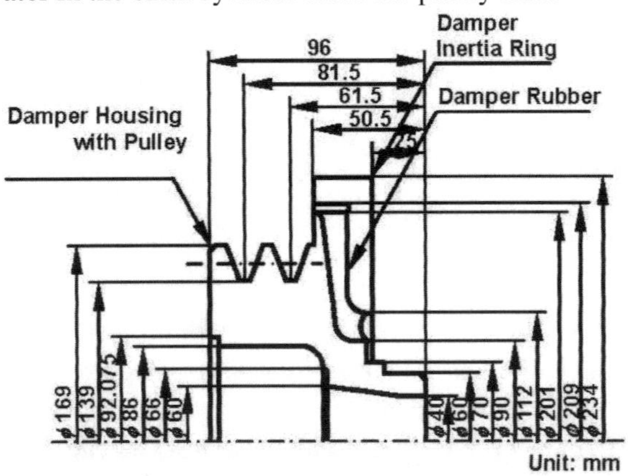

Fig. 1. Dimensions and Shape of the Test Rubber Damper Pulley

4. Dynamic Characteristics of Rubber Parts of Damper Pulleys

4.1 Measured Torsional Vibration Waveforms. The measured waveform of the pulley is the waveform in which the 6th-order amplitude is amplified due to the measurement at the 6th-order resonant point. On the other hand, the measured waveform of the damper inertia ring via the rubber part shows a slightly different tendency, in which the 6th-order amplitude is not particularly amplified and, in addition, the 3rd- and 4.5th-order amplitudes are large.

4.2 Measured Amplitude Curves of Angular Displacement. Figure 2 shows the measured amplitude curves of angular displacements at the pulley and the damper inertia ring of the crankshaft with the B-type damper pulley. These figures show merely the main-order amplitude curves of angular displacements. The large 6-th order resonant point occurs at 2524 [r/min] and its resonant amplitude is 14.9×10^{-3} [rad] in the case of installing no damper pulley. Since the tuning ratio of the B-type damper pulley is not optimum, the 6-th order amplitude is not greatly reduced.

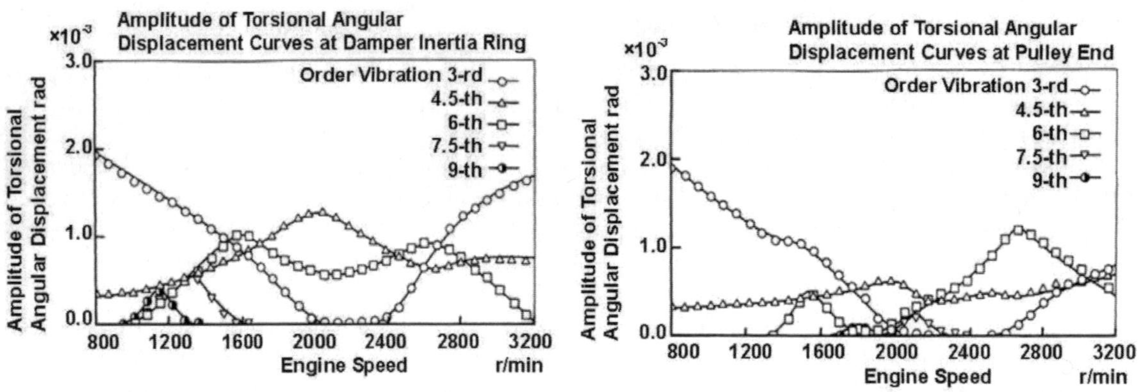

[a] Amplitude Curves of Inertia Ring **[b] Amplitude Curves of Damper Pulley**

Fig. 2. Measured Amplitude Curves of Angular Displacement [with B-Type Rubber Damper Pulley]

4.3 Numerical Calculation of Dynamic Characteristic Values from Measured Waveform. The values of dynamic torsional stiffness and damping coefficient of damper rubber parts can be obtained from the harmonically analyzed results of waveforms measured at the damper inertia ring and the pulley. The equation of motion at the damper inertia ring is

$$I_{RDP}\cdot\frac{d^2\theta_{RDP}}{dt^2}+C_{RDP}\left(\frac{d\theta_{RDP}}{dt}-\frac{d\theta_{PUL}}{dt}\right)+K_{RDP}\left(\theta_{RDP}-\theta_{PUL}\right)=0,$$
$$I_{RDP}\cdot\ddot{\theta}_{RDP}+C_{RDP}\left(\dot{\theta}_{RDP}-\dot{\theta}_{PUL}\right)+K_{RDP}\left(\theta_{RDP}-\theta_{PUL}\right)=0 \tag{1}$$

where $\theta_{RDP}=\theta_{RDP,O}\cdot e^{j(\omega t-\phi)}$ and $\theta_{PUL}=\theta_{PUL,O}\cdot e^{j\omega t}$. The following Eq. (2) can be obtained by rearranging the expression of Eq. (1) by substituting the above-mentioned relational expression into Eq. (1) and rearranging;

$$K_{RDP}=\frac{I_{RDP}\cdot\omega_{RDP}^2\cdot M_{RDP}\left(M_{RDP}-\cos\phi_{RDP}\right)}{M_{RDP}^2+1-2\cdot M_{RDP}\cdot\cos\phi_{RDP}},\quad C_{RDP}=\frac{I_{RDP}\cdot\omega_{RDP}^2\cdot M_{RDP}\cdot\sin\phi_{RDP}}{M_{RDP}^2+1-2\cdot M_{RDP}\cdot\cos\phi_{RDP}} \tag{2}$$

In Eq. (2), the values of amplitude ratio M_{RDP} and phase angle ϕ_{RDP} can be obtained by analyzing harmonically the waveforms measured at the damper inertia ring and the pulley, where the values of I_{RDP}, ω_{RDP} are known. Therefore, the values of dynamic torsional stiffness K_{RDP} and damping coefficient C_{RDP} can be determined from Eq. (2). The dynamic characteristics of the vibration isolator rubber depend generally on (1) temperature, (2) frequency, and (3) average strain and strain amplitude effects on the basis of same shape and material. To reveal the effect of one factor among the above-mentioned factors on the dynamic characteristics, it is necessary to keep the values of the other factors constant in the experiment. In our experiments, the surface temperature of the rubber part was kept at 313 [K], but the other factors could not be controlled due to the vibration characteristics of the crankshaft.

IOP Conf. Series: Materials Science and Engineering **517** (2019) 012003 doi:10.1088/1757-899X/517/1/012003

In full consideration of these conditions of the other factors, strain rate γ_{RDP}, which is the product of strain amplitude and angular frequency, is defined by the following equation.

$$\theta_{RDP} = \theta_{Re\,l} \cdot \omega_{RDP} \cdot \rho_{RDP} / b_{RDP} \tag{3}$$

$$\rho_{RDP} = \frac{3 \cdot \left(r_{RDP,2}^{4} - r_{RDP,1}^{4}\right)}{4 \cdot \left(r_{RDP,2}^{3} - r_{RDP,1}^{3}\right)} \tag{4}$$

Equation (4) is obtained for the uniform, hollow, circular cross section under the condition that the torsional moment (in such a case that the shearing stress in the representative radius ρ_{RDP} is uniformly distributed over the entire cross section) is equal to that (in the case that shearing stress distribution $\tau_{\rho_{RDP}} = G_{RDP} \cdot \rho_{RDP} \cdot C_{RDP,x}$ caused by the simple shearing deformation in the circumferential direction. Figure 3 shows the relationship between absolute torsional stiffness and strain rate. Figure 4 illustrates the relationship between loss factor and strain rate. Both Figs. 3 and 4 are aimed at expressing the absolute torsional stiffness and the loss factor as a function of the strain rate on the condition that the nonlinear dynamic characteristic values of the damper rubber part are assumed to be approximately expressed by linear equations. The values of the dynamic torsional stiffness K_{RDP} and the damping coefficient C_{RDP} (or $C_{RDP} \cdot \omega_{RDP}$), which are necessary as input data for the simulation analysis in section 4, are arranged in the same way as shown in Figs. 3 and 5, respectively. The value of absolute torsional stiffness can be determined by $\left|K_{RDP}^{*}\right| = \sqrt{(K_{RDP})^2 + (C_{RDP} \cdot \omega_{RDP})^2}$, but the following expression applies under the condition that the value of $C_{RDP} \cdot \omega_{RDP}$ is fairly small in comparison with the value of K_{RDP}; $\left|K_{RDP}^{*}\right| = K_{RDP}$. Therefore, the values of $\left|K_{RDP}^{*}\right|$ and K_{RDP} approximately agree in the figure. The curves obtained by the curve-fitting method are illustrated in Figs. 3, 4 and 5. The values of the correlation coefficients against the curves of Dampers A and B are in the regions of 0.82-0.89 and 0. 86-0. 94, respectively.

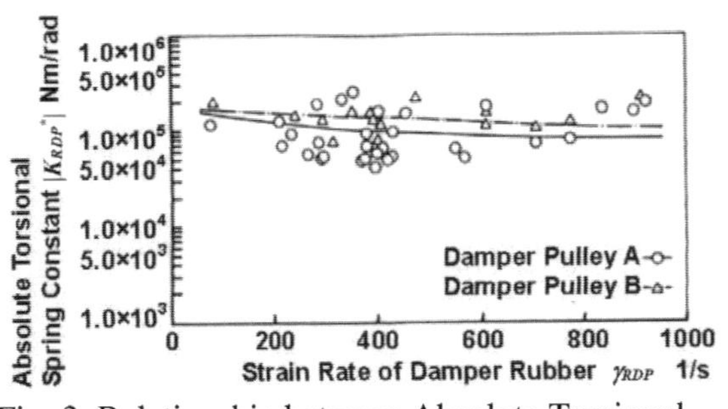

Fig. 3. Relationship between Absolute Torsional Stiffness [Dynamic Torsional Stiffness] and Strain Rate.

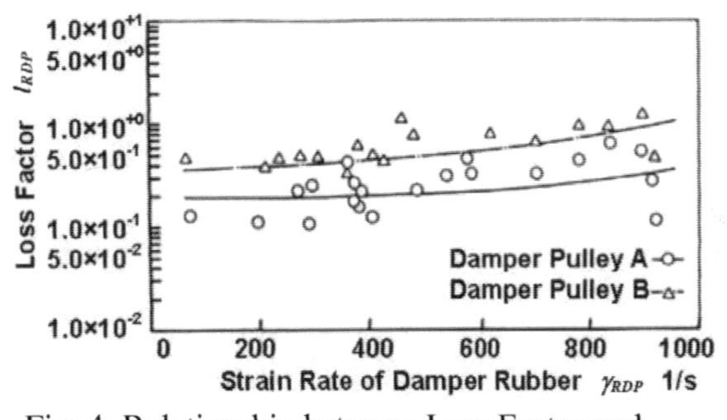

Fig. 4. Relationship between Loss Factor and Strain Rate.

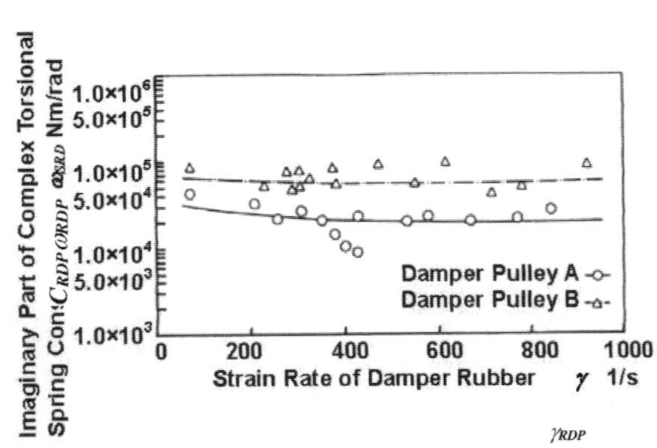

Fig. 5. Relationship between Imaginary Part of Complex Torsional Stiffness and Strain Rate.

5. Analytical Investigation of Torsional Vibration Characteristics of Engine Crankshaft with Damper Pulley

5.1 Input Data Necessary for Numerical Calculation of Torsional Vibration Waveforms. Figure 6 shows the equivalent torsional vibration system of a crankshaft with a rubber damper pulley replaced according to the analytical method described in section 4.1. The rubber part of the damper pulley is replaced with a Voight model. Since the simulation program can yield the torsional vibration waveform for a given engine speed, the value of the strain rate is numerical calculated using the measured dominant-order amplitude of angular displacement at a given engine speed. Then, the values of the dynamic torsional stiffness K_{RDP} and damping coefficient C_{RDP} of the rubber part can be obtained by referring to Figs. 3 and 5, respectively. These obtained values of K_{RDP} and C_{RDP} are used as the input data for the numerical calculation. Since the curve-fitted curves in Figs. 3 and 5 are representative of all the experimental data, the obtained values of K_{RDP} and C_{RDP} can be regarded as the mean values.

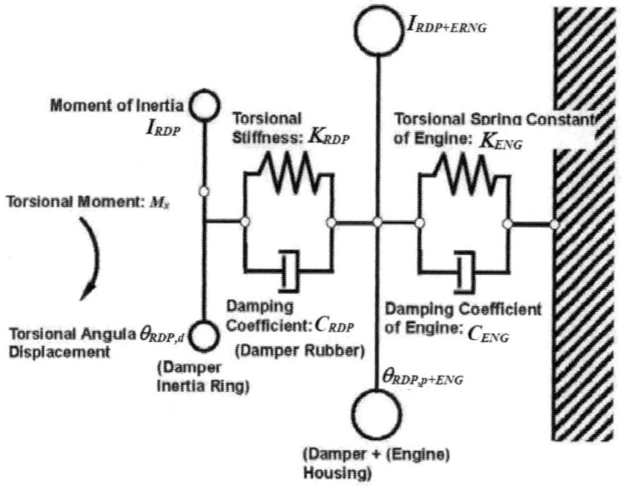

Fig. 6. Equivalent Torsional Vibration System of an Engine Crankshaft System with a Rubber Damper Pulley

5.2 Numerical Calculated Results and Considerations. Figure 7 shows the main-order amplitude curves of the angular displacements at the damper inertia ring and the pulley obtained by analyzing harmonically the numerical calculated torsional vibration waveforms of the crankshaft with the rubber damper pulley. The numerical calculated amplitude curves in this figure correspond to the measured amplitude curves of the angular displacements in Fig. 2. As compared with the experimental results, these calculated results contain slight error, but this degree of error is allowable in practice.

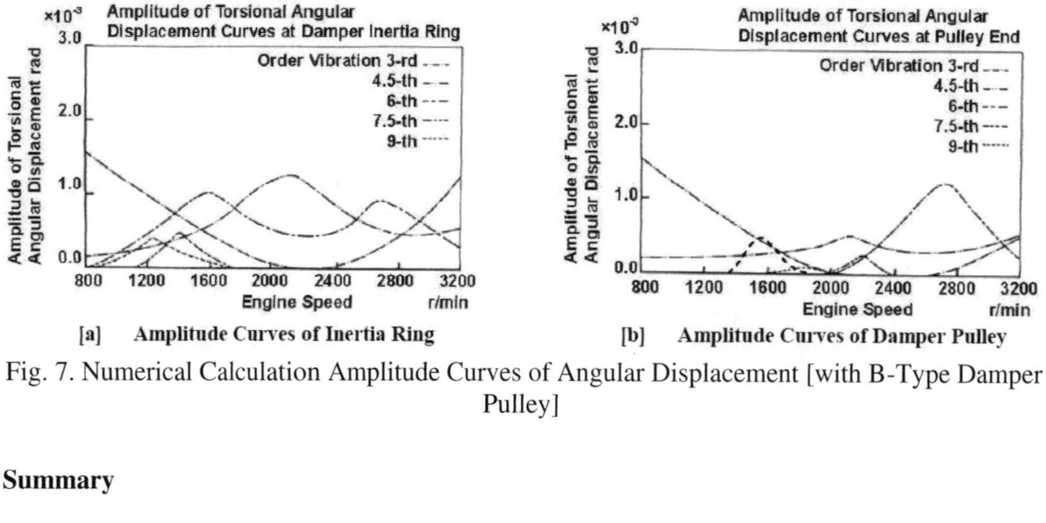

Fig. 7. Numerical Calculation Amplitude Curves of Angular Displacement [with B-Type Damper Pulley]

6. Summary

Accurate estimation of the dynamic characteristics of the rubber part of the damper pulley are problems in the numerical calculation of torsional vibration of a crankshaft with a rubber damper pulley. The results of investigation of these characteristics from experimental and analytical viewpoints are as follows:

(1) We expressed the absolute torsional stiffness and damping of the rubber part of the damper pulley as a function of the strain rate under the conditions of the same kind rubber material, constant temperature and invariable shape factor.

7. References

[1] Yasuhiro Honda, Tomoaki Kodama, Katsuhiko Wakabayashi : Relationships between Rubber Shapes and Dynamic Characteristics of some Torsional Vibration Rubber Dampers for Diesel Engines, *The 15th Pacific Automotive Engineering Conference – APAC15*, No.0320, (2009), p. 1-8.

[2] Yasuhiro Honda, Katsuhiko Wakabayashi, Tomoaki Kodama, Hiroshi Okamura : An Experimental Study of Dynamic Properties of Rubber Specimens for a Crankshaft Torsional Vibration of Automobiles, *ASME 1999 International Design Engineering Technical Conference and The Computers and Information in Engineering Conference, 17th Biennial Conference on Mechanical Vibration and Noise Symposium on Dynamics and Vibration of Machine Systems*, Session VIB-00037: Dynamics and Vibration of Machine System 02, DETC99/VIB-08125, (1999), p. 1-14.

[3] Yasuhiro Honda, Katsuhiko Wakabayashi, Tomoaki Kodama, Shinkichi Hama, Shoichi Iwamoto : An Experimental Study on Dynamic Properties of Rubber Test Specimens for Design of Shear-type Torsional Vibration Dampers, *The Ninth International Pacific Conference on Automotive Engineering*, Volume 2, No. 971400 (Abstract Code: 00088), (1997), p. 217-222.

[4] Tomoaki Kodama, Yasuhiro Honda, Katsuhiko Wakabayashi, Shoichi Iwamoto : A Calculation Method for Torsional Vibration of a Crankshafting System with a Conventional Rubber Damper by Considering Rubber Form, *SAE 1996 International Congress and Exposition*, SAE Technical Paper Series No. 960060, (1996), p. 103-121.

[5] George Nerubenko: Tuned Torsional Vibration Damper, *SAE 2013 Noise and Vibration Conference and Exhibition*, SAE Technical Paper 2013-01-1897, DOI: https://doi.org/10.4271/2013-01-1897, (2013), p. 1-8.

[6] Mohammad Hijawi: Dual Torsional Vibration Damper, *SAE 2012 World Congress and Exhibition*, 2012-01-0065, DOI: https://doi.org/10.4271/2012-01-0065, (2012), p. 1-9.

[7] George Nerubenko: Engine Noise Reduction Using Self-Tuning Torsional Vibration Damper, *SAE 2016 World Congress and Exhibition*, SAE Technical Paper 2016-01-1063, DOI: https://doi.org/10.4271/2016-01-1063, (2016), p. 1-8.

[8] Jaroslaw Pankiewicz and Bogumil Chilinski: Model Aided Design of Tuned Rubber TVD, *Journal of KONES Powertrain and Transport*, Vol. 23, No. 4, DOI: 5604/12314005.1217250, (2016), p. 367-374.

[9] Hanmant. S. Tamkhade and G. S. Kondhalkar: Theoretical and Experimental Validation of Viscous Torsional Damper on Turbocharged Inline Six Cylinder Engine, *International Journal of Innovations in Engineering Research and Technology*, Vol. 4, Issue 2, (2017), p. 59-72.

[10] John Licari, Carlos E. Ugalde, Janaka B. Ekanayake and Nick Jenkins: Comparison of the Performance and Stability of Two Torsional Vibration Dampers for Variable-Speed Wind Turbines, *Window Energy*, John Wiley & Sons, No. 18, (2015), p. 1545-1559.

[11] Jing Sheng, Meng-da He and Mei-hua Yao: Simulation Study on the Performance of a Rubber-type Torsional Vibration Damper, *2017 2nd International Conference on Applied Mechatronics Engineering*, (2017), p. 79-84.

[12] Tieshn Zhang and Mingsong Li: A Design Study on the Torsional Vibration of the Vehicle Powertrain, *5th International Conference on Advanced Engineering Materials and Technology*, Atlantis Press, (2015), p. 429-433.

[13] Lionel Manin, Regis Dufour and Sebastien Schultz: Pulley Torsional Vibraion Damper Characterization, *Mechanics & Industry*, AFM, EDP Sciences, Vol. 14, DOI: 10.1051/meca/2013057, (2013), p. 151-155.

IOP Conf. Series: Materials Science and Engineering **517** (2019) 012004 doi:10.1088/1757-899X/517/1/012004

Automatic continuous manufacturing method of lamp holder vertical injection molding machine

Liu Yu[1], Chen Jing[2] and Chen hao[3]

1 Postgraduate, Wuhan University of Technology, Wuhan, China
2 Professor, Wuhan University of Technology, Wuhan, China
3 Postgraduate, Wuhan University of Technology, Wuhan, China
E-mail: liuyuflhs@163.com

Abstract. The manufacting of lamp holder usually requires the coordination of the handmade operator, which is slow and inefficient. So, the automatic continuous manufacting method of the lamp holder vertical injection molding machine is discussed. This method is based on the sorting, feeding, mold injection and blanking technology of lamp holder. The processing technology and process flow of the lamp holder are introduced, and the collaborative control method of the lamp holder is expounded. This method improves the precision and reliability of the lamp holder injection molding and the degree of automation, and reduces the security hidden danger.

1. Introduction

The lamp holder is generally divided into plastic lamp holder, bakelite lamp holder, ceramic lamp holder and metal lamp holder etc. The ceramic lamp is easily broken, so transportation is not convenient .Bakelite lamp holder is non-recyclable, and the cost is expensive. Plastic lamp holder is more popular because the heat-resistant temperature of it is higher than 240 degrees and difficult to be damaged. Now, the world-famous plastic lamp holder in the global market is generally provided by BJB of Germany and ARDITI of Italy. There are many manufacturers of plastic lamp holder in China, but quite well known, it can't match many foreign manufacturers [1].

The plastic lamp holder is usually molded by injection molding machine and corresponding mold, its raw material is generally PET or PBT and LCP.Usually, the injection material of the plastic lamp holder is injected into the top of the metal lamp holder thus get the finished product [2]. Currently, most of the lamp holder injection molding machines only have the function of injection molding, however sorting , feeding and blanking of lamp holder usually done with the cooperation of workers which is slower, poor efficiency , and easy to cause security risks[3] [4].

In view of the above problems, an automatic continuous manufacturing method of the lamp holder used to vertical injection molding machine with high speed, high precision, high efficiency and compound function is studied.

2. Automatic continuous manufacturing process of lamp holder vertical injection molding machine

The automatic continuous manufacturing method of lamp holder vertical injection molding machine is consist of sorting, automatic feeding, molding, automatic blanking, and its process flow is shown in Fig 1.

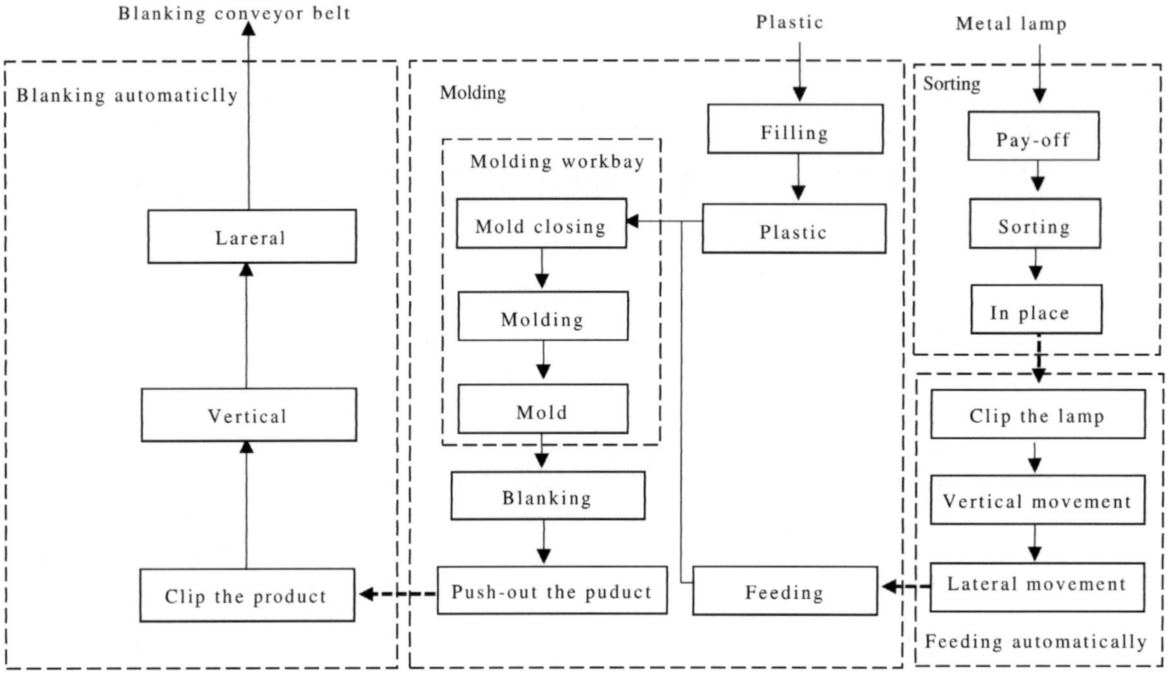

Fig1. Workflow chart of automatic continuous manufacting of lamp holder vertical injection molding machine

2.1. Sorting

In traditional, the separation and sorting of the metal lamp holder is usually manually. Workers separate the metal lamp holders which are clinging together, it's inefficient and high-risk.

So, the automatic sorting module is used, aim to separate and sort the lamp holders automatically .The automatic sorting module consists of the material slot, the conveyor belt,the vibrating feed plate, the supporting platform.The material slot is used to place the metal lamp holders. The lamp holders are transmitted to the vibrating feed plate by conveyor belt after receive the sort instruction from controller. The vibrating feed plate divide the clinging lamp holders and sort them. In sorting process, if the direction of the lamp holder is incorrect, it fall into the trough waiting for the next sorting. The sorted lamp holders are sent to the supporting platform through a conveyor belt and waiting in place. When the feeding manipulator have taken away the lamp holders in place, the next sorting is carried out, so it is repeated.

2.2. Feeding automatically

In traditional, the feeding is manually, workers put the lamp holders on the feeding workbay by hands, this way is not only inefficient and exist a big security risk.

So, the feeding manipulator is used for replacing workers to put the lamp holders on feeding workbay. The feeding manipulator consists of the clamp of feeding manipulator, the vertical lift frame, the horizontal guide rail. The horizontal movement of the clamp is driven by a servo motor, and the loosening of the vertical movement and the fixture is pneumatically driven.

The feeding manipulator receive the signal from controller which allow to feeding. Then after vertical lifting and horizontal movement ,manipulator get to above of the support platform, the clamp of manipulator clip the lamp holders .And then after the vertical and horizontal movement,manipulator get to the feeding workbay on main body of injection molding machine .Then the fixture release, and place the lamp holders in the corresponding position, feeding is completed. Then Feeding manipulator give a signal to inform the injection molding machine that it can take the next step.

2.3. Molding

The molding is completed by injection molding machine and corresponding mold. The injection molding machine consists of the feeding funnel, the melting tube, the working turntable (feeding workbay, molding workbay, blanking workbay), and the ejection device. The mold is composed of an upper mold installed on the mold clamping device and three lower molds installed on the corresponding workstations.The molding mainly complete the plastic filling, plastic melt, injection molding, mold opening, product pushing. Thereby the metal lamp holders to be finished products.

Specifically, after the injection molding machine receives the feeding completion instruction sent by the controller, the working turntable transfers the lamp holders from the feeding workbay to the molding workbay at where melted plastic will be injected into mold.

After forming, cooling, open the mold. The Injection molding is completed, then the working turntable will turn the molded lamp holders to the blanking workbay where the molded lamp holders will be pushed out by the ejection device .In this moment the injection molding machine send a signal to the blanking manipulator to allow blanking, then waiting for the blanking.

2.4. Blanking automatically

The traditional process of removing the finished product is manually, and the worker takes out the finished product of the lamp at the blanking workbay. It has the same problems of inefficient and unsafe as the manual feeding.

So, the blanking manipulator is used for replacing workers to remove the lamp holders on blanking workbay. The storcture of the blanking manipulator is the same as the feeding manipulator. The blanking manipulator receive the signal from controller which allow to blank. After vertical lifting and horizontal movement, manipulator get to above of the blanking workbay on Injection molding machine .The clamp of manipulator clip the lamp finished product .And then after the vertical and horizontal movement, manipulator get to the above of reclaimer conveyor belt .Then the clamp release, place the lamp holders in the corresponding position.Then blanking manipulator send a signal to inform the jection molding machine that it can take the next step.

3. Control strategy

The structure of the injection moulding machine results in the discrete distribution of its electrical components and the local concentration of the main control box. The long-distance transmission of IO signals is an important factor affecting accuracy. In this paper, the distributed with centralized control is used to collect and process analog signals in situ and transmit data to the main controller in the form of digital signals through bus. This control method can avoid the interference of analog signals in the long-distance transmission process, thus realizing IO remote distribution. The distributed with centralized control systems is characterized by flexibility, reliability, coordination, and ease of maintenance.Failure of a part of the system will not result in the loss of other functions of the system, and the troubleshooting can be done without affecting the entire system [5].

3.1. Control system

Here, a distributed system with centralized control is adopted.The system consists of automatic sorting module, feeding manipulator, Injection molding machine body, blanking manipulator and ejection device. Each part has its own controllers. At the same time, each part is controlled centrally by the central control unit. Each module transmits information to the central control unit through the bus. The central control unit can receive the status information of each module in real time, and coordinate and control each module according to the current status. [6] [7]. The signal control is used here instead of time control. That is, only when receive the signal then take to start the next action, rather than a regular action. The benefits of doing so are more real-time and more capable of dealing with emergencies. At the same time, it avoids the time deviation which is easy to exist in timing action.

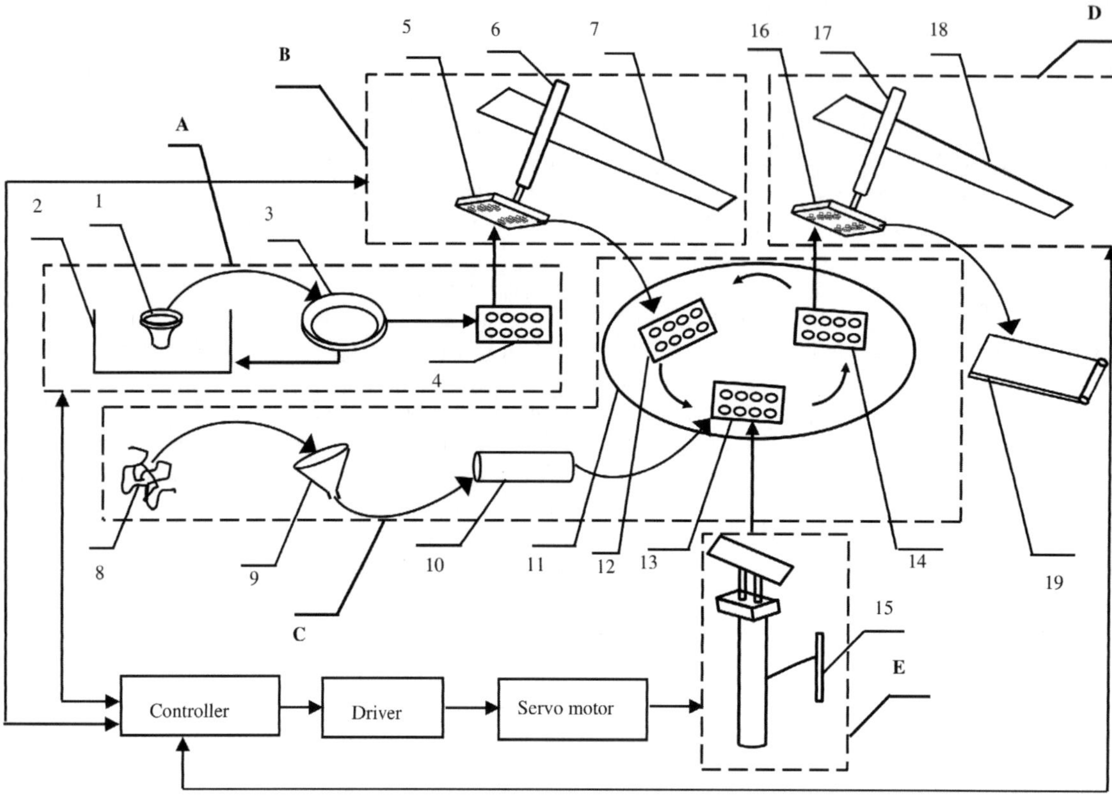

Fig 2. Synchronize workflow chart of automatic continuous manufacting of lamp holder

A. Automatic sorting module; B. Feeding manipulator; C. Injection molding machine body; D. Blanking manipulator; E. Ejection device; 1. Lamp holder; 2. Material slot; 3. Vibrating feed plate; 4. Supporting platform; 5. Clamp of feeding manipulator; 6. Vertical lift frame; 7. Horizontal guide rail; 8. Plastic; 9. Feeding funnel; 10. Melting tube; 11. Working turntable; 12. Feeding workbay; 13. Molding workbay; 14.Blanking workbay; 15. Displacement sensor; 16. Clamp of blanking manipulator; 17. Vertical lift frame; 18. Horizontal guide rail; 19. Reclaimer conveyor belt.

3.2. Synchronize workflow

The four processes of sorting, feeding automatically, molding, blanking automatically can run independently, and also works synergeticlly.The synchronize workflow is shown in Fig2.

3.2.1. Sorting and feeding simultaneously .The automatic sorting module sorts the lamp holders, after that the feeding manipulator take the lamp holders away. Then the automatic sorting module start the next sorting process, at the same time the feeding manipulator take the lamp holders to the feeding workbay on injection molding machine.The two processes work simultaneously.

*3.2.2 Feeding and molding simultaneously.*After the feeding manipulator finish a process, the injection molding machine then begins workbay converting, injection molding, mold opening and product pushing. At the same time the feeding manipulator move to the sorting module for the next reclaimer.The two processes work simultaneously.

*3.2.3 Molding and blanking simultaneously.*The blanking manipulator remove the finished product from blanking workbay on injection molding machine to reclaimer conveyor belt .At the same time, the next batch of lamp holders are molding at molding workbay on injection molding machine.The two processes work simultaneously. The two processes work simultaneously.

*3.2.4. Blanking and feeding simultaneously.*The blanking manipulator remove the finished product from blanking workbay, at the same time the feeding manipulator take the lamp holders to the feeding workbay. The two processes work simultaneously

4. Conclusion

In this paper, The automatic continuous manufacturing method of the lamp holder vertical injection molding machine is designed to solve the problems in most of the domestic injection molding processes.This method ameliorate the problems of the simple function, low injection precision, low speed, high-cost, labor intensity of workers is great, low degree of automation and production efficiency, and poor market competitiveness and so on. This method is applied to the vertical injection molding machine to realize the automatic continuous manufacturing of the lamp holder. This method enhances the degree of automation of the injection molding machine, improves the accuracy and reliability of the injection molding machine, liberates the labor, improves the production efficiency, eliminates the manual feeding and blanking work mode and reduces the potential safety hazard.

5. References

[1] Xu Baolin. Design of Control System for a Large-Size Injection Molding Machine [J]. Modern Plastics Processing And Applications，2005，17 (6):42-45.

[2] Yu Fufang，Cai Shaohan，Chen Hongyi，Qiu Yiqian. Design of Injection Mold for Energy-Saving Lamp Holder [J]. Engineering Plastics Applications, 2011, 39(1):74-77.

[3] Zhang Yougen. Scientific Development and Innovation of Functional Injection Molding Machine [J].CHINA RUBBER/PLASTICS TECHNOLOGY AND EQUIPMENT (Plastic), 2015, 41(16):10-33.

[4] Zhang You Gen. Discussion About The Character And Developing Direction Of Plastic Injection Molding Machine For China [J]. Shanghai Plastics, 2004, 125: 4-8.

[5] Xiang Peng. Distributed Control System of All electric IMM [J]. Machinery and electronics, 2012, 5:33-35.

[6] Ahmed, Engr Syed Faiz. A new approach in Industrial automation application" embedded system design for Injection Molding Machine"[J]. Multitopic Conference, 2007.INMIC 2007. IEEE International: 1-5.

[7] Yan Wei. Application of distributed control system in large injection molding machine [J]. Light Industry Machinery, 2013, 31:105.

6. Acknowledgments

This work was financially supported by 2017 Zhongshan Science and Technology Development (Seco nd Batch) Projects (Granted No: 2017A1031, Project Name: Manufacture and Industrialization of Hig h-performance Integrated Vertical Injection Molding Machine).

IOP Conf. Series: Materials Science and Engineering **517** (2019) 012005 doi:10.1088/1757-899X/517/1/012005

Model-based multi-critical optimisation of combustion engine fuel consumption and emissions

Xiaoming Wang[1], Shujun Zhang[1] and Kemal Bechkoum[1]

[1] School of Computing and Technology, University of Gloucestershire, UK
E-mail: xiaomingwang@connect.glos.ac.uk

Abstract. The combustion engine is a typical nonlinear multi-input multi-output (MIMO) system with strong couplings, actuator constraints, and fast dynamics. This paper addresses a model-based multi-critical optimisation approach in diesel engines, which allows to improve emission performance and to provide a reference for the design and optimisation of the diesel engine system. The first part of this paper introduces a data-based modelling method that appears particularly suitable for emission modelling. The Design of Experiments (DoE) method helps to generate and collect the required measurement for data-based modelling in a short time, despite the increasing number of manipulated variables. The second part establishes a new model-based multi-critical optimisation approach that supports the optimisation of fuel consumption and emissions based on engine models. This proposed model-based framework consists of system identification and multi-critical optimisation. This framework has the ability to achieve the fast and precise solving of multi-critical optimisation problem and is suitable for implementation in the engine control unit. The experiment results illustrate that the model-based multi-critical optimisation significantly improves the engine exhaust emissions and fuel consumption against the original ECU.

1. Introduction

The diesel engine plays a dominant role in heavy-duty vehicles, agricultural machinery, engineering machinery, and other fields due to its high energy efficiency, strong driving performance, and good economic characteristics. Shown in Figure 1 is a common configuration in many modern diesel engines for a power vehicle system, especially when high performance is required. However, the primary problem for the diesel engine is its emissions of nitrogen oxides (NOx) and particulate matter (PM; also called OPAC when measured as opacity) [1]. With increasingly stringent emission regulations, the question of how to improve the economy and emissions of diesel engines has become an important issue for engine development [2]. For diesel engines, this pursuit involves using control optimisation as well as exhaust gas recirculation (EGR) technology to reduce NOx emissions, and employing after-treatment technology, such as three-ways catalytic, diesel particulate soot filters, and other methods, to reduce its particulate emissions [3]. Furthermore, the engine turbocharger technology is an important means to improve engine performance. Variable-geometry turbochargers (VGTs) are a family of turbochargers usually designed to allow the ratio of Area/Radius (A/R) of the turbo to be adapted as engine operations change [4]. Variable-geometry turbochargers can reduce the turbine vane's variable angle opening when the engine is running at low speed, thus increasing the exhaust gas pressure and flow rate, improving the exhaust energy efficiency, and enhancing the low speed torque performance. The turbine vane's variable angle opening can be adjusted during high-speed operation, so that it works within the high-efficiency area of the turbocharger. When combined with the EGR system, one can adjust the opening of VGT vanes to improve the pressure difference between the turbine and EGR system, which could further improve the EGR rate and reduce NOx emissions.

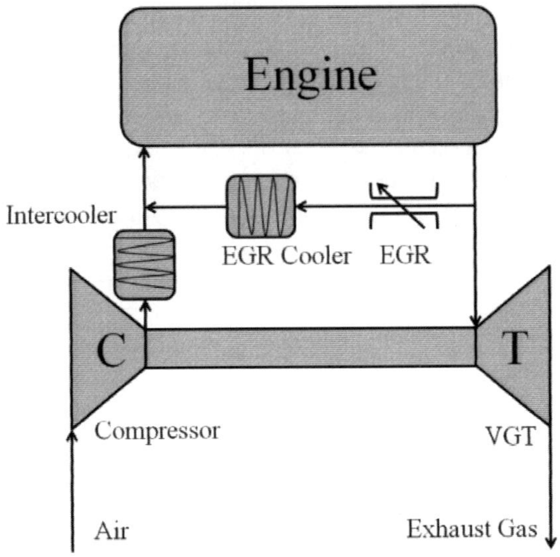

Figure 1. Turbocharged Diesel Engine Equipped with EGR

Therefore, the classic manipulated variables of injection volume and the injection process are joined by further manipulated variables, such as EGR, VGT, injection pressure in common rail systems, variable valve train (VVT), and injection volume modulation. These control variables affect torque, fuel consumption, and emissions. The cross-coupling effect of control variables creates a complex nonlinear multivariable system. The variety of control options and their interactions make it increasingly difficult for engine designers to find an optimal engine setting, and the number of realisable characteristics in modern engine controls rise sharply. For some time, the traditional test-bench approach, which is limited to a stationary assignment of the inputs and outputs variables listed above, has not met all requirements [5]. Hence, model-based methods are required to develop engine control systems further that permit the static and dynamic behaviour of internal combustion engines to be exactly determined with the help of a more precise mathematical simulation. This paper presents an integrated method based on emission modelling to optimise the consumption and emission behaviour of internal combustion engines.

In the first part of the paper, the modelling of emissions will be explained in detail. With the help of the DoE approach, the required measurement data for data-based modelling of emissions can be collected in a short time despite the increasing number of manipulated variables. The validation results of the emission models are then presented and discussed. The second part of this paper describes how the presented emission models form the basis of a model-based optimisation of fuel consumption and emissions.

2. Data-based modelling of emissions

In many real-world situations, it is too difficult to describe a system using known physical laws. With engine exhaust emissions, for example, the system identification method can be used to perform data-based modelling. The goal of system identification is to estimate a model of a system based on observed input-output data [6]. System identification is useful when the only available information from a system is input and output data. The procedure to determine a model of a dynamic system from observed input-output data involves three components: 1) the input-output data; 2) the model structure- possible candidate models; and 3) the identification method (some criterion to select a model in the set, given the information in the data).

Identifying the engine exhaust emission model uses the data set collected from the engine test bench. Figure2 presents the emission model structure, which includes all the inputs and outputs for identification. To obtain this measurement data, a DoE plan was applied to the engine. The input signals are mf: injected fuel mass in mg/cycle; n: engine speed in rpm; MAP: manifold absolute pressure in mbar; MAF: manifold air flow in mg/cycle. The output signals are NOx in ppm and OPAC in %. Figure 3 illustrates the results of the DoE data in 3D. The duration of the measurements for the training data sets was around 0.5 seconds at every operating point, whereby the control variables were changed abruptly in a region around the series setting. The sampling time was 100ms.

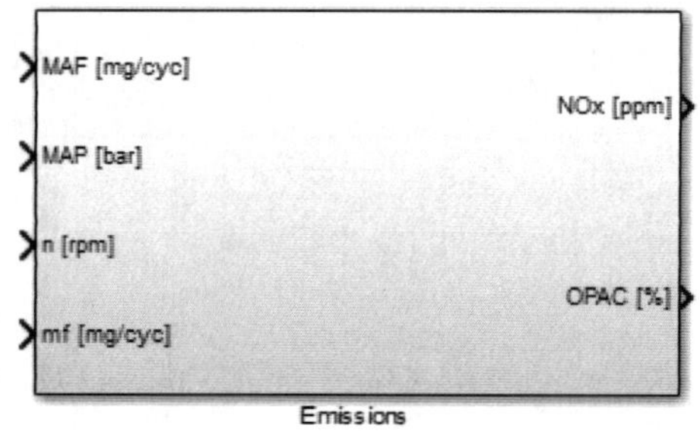

Figure 2. Emission Simulink Model

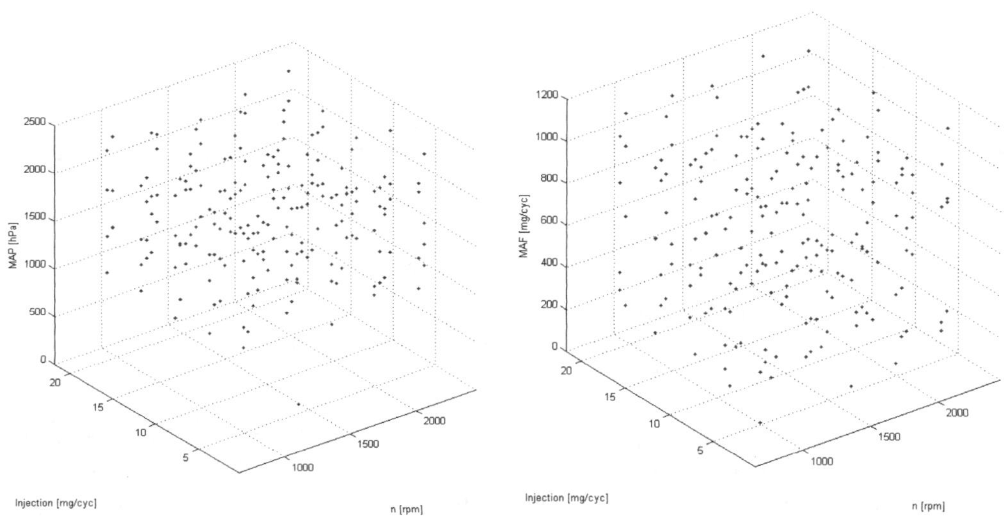

Figure 3. DoE-Data (Engine Injection, Engine Speed, MAP, and MAF) in 3D

System identification can be performed from these measurements. The engine emission model is identified in one step, with a MIMO structure to catch the interactions between all inputs and outputs. The system identification toolbox from Matlab enables to estimate models of different types and orders. To perform the validation, Figure 4 shows that the NLARX model estimations yield a reasonably close response to the original measurement data. The average error of NLARX model was between 5% and 10%. Thus, the identified model is able to represent the real behaviour of the engine emissions in the given operating range. So far, the output variables (NOx and OPAC) were modelled with the engine control variables as model inputs. This emission model is further connected to a mean-value engine model which was modelled in a previous work [7]. Alternatively, internal variables from the measurement dataset, such as MAF and MAP can be used as input variables for the models. The MAF and MAP used as input are useful when the optimisation aim introduced in the second part is not the calculation of control characteristics for the manipulated variables but when calculating command variable mapping characteristics for secondary control (VGT or EGR) [7].

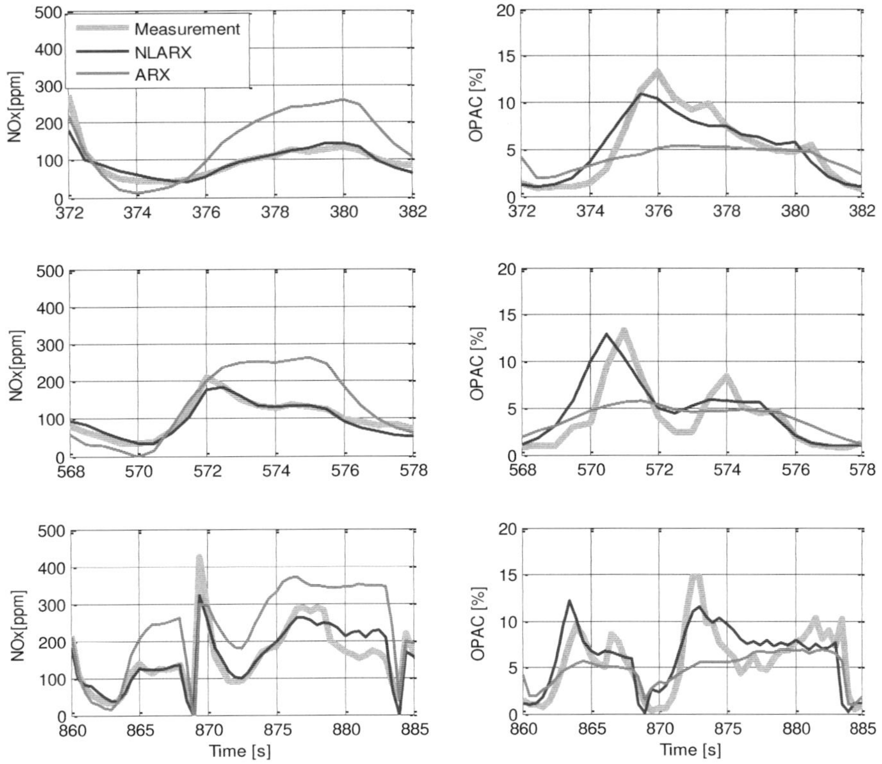

Figure 4. Comparison of Measurement, NLARX and ARX Model

3. Model-based multi-critical optimisation of fuel consumption and emissions

With engine optimisation, the exhaust emissions must be restricted for the duration of an operating cycle to the legally permissible limit for each emission component with minimal fuel consumption. These two requirements often contradict each other and reducing certain exhaust gas constituents typically entails an increase in consumption, which is why an appropriate compromise must be found. Experienced test engineers traditionally carry out optimisation based on test-bench conditions [7]. This procedure is time consuming and is impossible to implement when the number of manipulated variables rises. In this paper, the engine behaviour is presented with a mathematical model (mean-value engine model connected with emission model); offline optimisation can be performed on this model to determine the optimum manipulated variables [8].

Optimisation is based on cost function J (Eq. 1) to determine fuel consumption and emissions. Engine control variables should minimise this cost function through optimisation algorithms. Optimisation tasks frequently lead to a situation where a reduction (e.g. of an emission variable) is associated with an increase of other values, such as consumption. The model-based multi-critical optimisation is available for the optimisation of such 'opposite' outputs.

$$J = k_1(n, mf)(norm(NOx))^2 + k_2(n, mf)(norm(OPAC))^2 + k_3(n, mf)(norm(fuel))^2 \quad (1)$$

Table 1. Elements of cost function

Name	Weight	Target
NOx	k_1	Low NOx Emission
OPAC	k_2	Low OPAC Emission
Fuel	k_3	Low Fuel Consumption $_{\text{Equivalent Power}}$

In order to achieve comparable sensitivity of the cost function for the respective components, it is necessary to first normalise them within each operating point. To this end, for any reference values, the respective behaviour can be used in the series products. The weighting factors are defined in the offline application of this procedure by different value combinations that appear useful. To prevent inhibiting the drivability of the vehicle, the engine torque must be not smaller than a predetermined moment represents an additional condition. Optimisation for each operating point means that Eq. 1 must be solved by an optimisation method. For this purpose, the method SQP (sequential quadratic programming) was used [9], which is designed for non-linear, multivariable problems with constraints and, for example, as a function called 'fmincon' in MATLAB's Optimisation Toolbox.

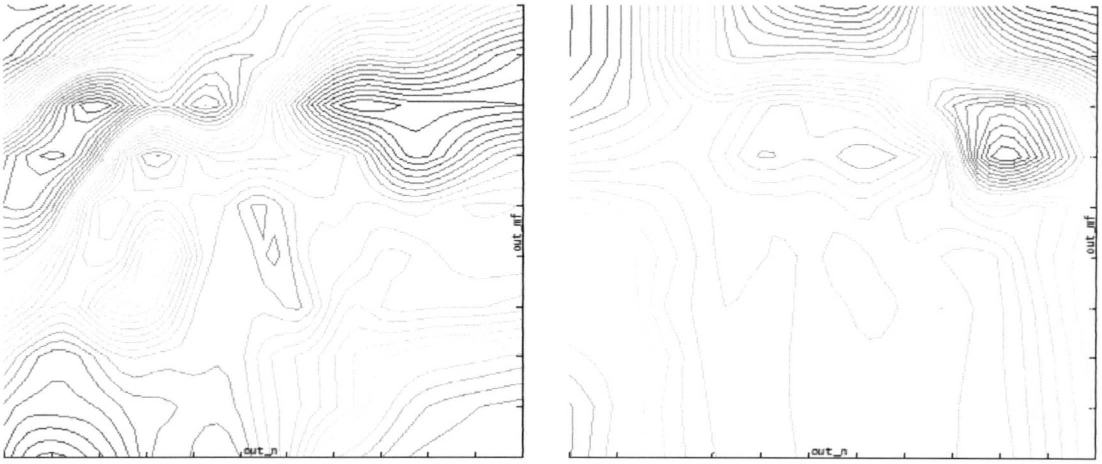

Figure 5. FTP-75 Cycle-Based Weighting Factors for OPAC (left) and NOx (right)

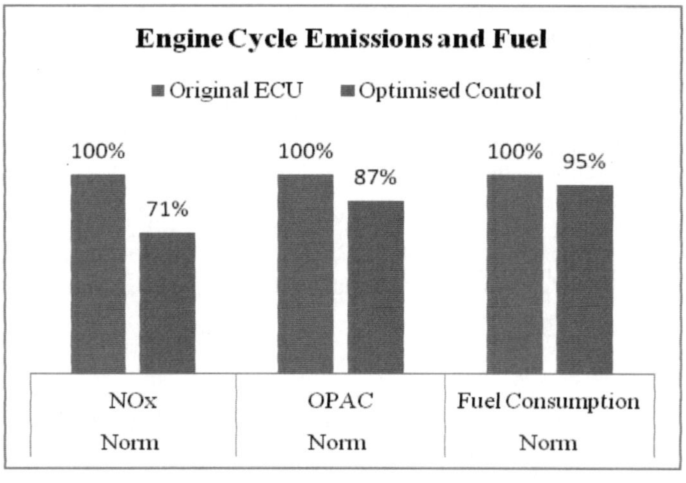

Figure 6. Comparison of Engine Cycle Emisisons and Fuel Consumption

After optimising, a control map is selected for driving that complies with the emission limit values for a particular driving cycle and that achieves the lowest consumption. If further optimised control maps for other weightings are defined in the control unit then a different emissions and fuel consumption composition can be set (e.g. depending on the driving situation while in operation). This cycle-based weighting, as shown in Figure 5 and positioning of the operation points, $k\ (n,\ mf)$, with the reference to the calculation rules can now be used for the global optimization. These optimised manipulated variable maps are further applied to the engine-air-path control. The results of the optimisation compared with the original ECU are shown in Figure 6. The simulation results showed that the NOx emission, OPAC and fuel consumption were decreased by 29%, 13% and 5% respectively against the original ECU.

4. Summary

To meet the increasing demands on modern diesel engines in terms of low consumption and minimal exhaust emissions, it is necessary to control the engine more optimally. The presented model-based multi-critical optimisation methods are carried out for multiple manipulated variables (Engine Injection, Engine Speed, MAP, and MAF) and outputs (fuel consumption and emissions) with cycle-based weighting factors in a relatively short time whereby any driving cycles and either stationary or transient engine models can be used. Optimisation, therefore, does not require a lengthy online engine test, but can be performed offline with simulation engine models. Overall, a model-based framework is thus available, which helps achieve optimisation of engine control within a relatively short amount of time.

5. References

[1] ACEA, "NEW PASSENGER CAR REGISTRATIONS EUROPEAN UNION", *ACEA*, Brussels, 2017.
[2] K. Reif, Diesel Engine Management: Systems and Components, *Berlin: Springer Vieweg*, 2014.
[3] Integer Research, "Market Overview Emissions Control in Europe," *Integer Research*, London, 2017.
[4] H. Nguyen-Schaefer, Aero and Vibroacoustic of Automotive Turbochargers, *Berlin: Springer*, 2013.
[5] H. Hochschwarzer, Automatische Motorparameteroptimierung, *Graz: Technische Universität Graz*, 1990.
[6] R. Johansson, System Modeling and Identification, *Chicago: Prentice Hall*, 1993.
[7] X. Wang, S. Zhang and K. Bechkoum, "Data-Based Gain-Scheduled Modeling and Nonlinear Control of Engine Intake and Exhaust System," *Advances in Engineering Research*, 2016.
[8] A. Mitterer and F. Zuber-Goos, "Modellgestütze Kennfeldoptimierung – Ein neuer Ansatz zur Steigerung der Effizienz in der Steuergeräteapplikation," *ATZ*, pp. 188-196, 2002.
[9] M. Hirsch, Identification of a Virtual Sensor Model for Diesel Egnine Emissions by Means of Optimal Input Design, *Linz: Trauner Verlag*, 2011.
[10] H. Sequenz, "Emission Modelling and Model-Based Optimisation of the Engine Control," *Technische Universität Darmstadt*, Darmstadt, 2013.

IOP Conf. Series: Materials Science and Engineering **517** (2019) 012006 doi:10.1088/1757-899X/517/1/012006

Research on Underbody Flattening of a Passenger Vehicle towards Aerodynamics Optimization

HE Xiaona[1,2], LI Tao[3], XI Chunfu [2], and Zhang Honghui[1]

1 State Key Lab. of Vehicle NVH and Safety Technology, Chongqing 401122, P. R. China
2 China Automotive Engineering Research Institute Co., Ltd., Chongqing 401122, P. R. China
3 Beijing New Energy Automobile Co., Ltd., Beijing 102606, P. R. China
E-mail: hexiaona@caeri.com.cn

Abstract. A CFD method for the full-scale external flow of a passenger vehicle is established. Based on the comparison between the simulation and wind tunnel test results, the reliability to use this model and method for the aerodynamic drag coefficient (Cd) and lift coefficient (Cl) prediction of the vehicle is verified. The trend of Cd and Cl for five models with different underbody flattening degree shows the significance of underbody flattening on the aerodynamic performance, with nearly 18% on the drag and over 30% on the lift coefficient optimization

1. Introduction
CFD is an effective and widely used method for the automotive aerodynamics, and has achieved the transition from the theoretical research into the practical engineering [1]. Using commercial software STAR-CCM+, full-scale models and the external flow field of a passenger vehicle was established in the literatures [2-4] to verify the effectiveness of CFD method by the contrast between the calculation and wind tunnel test, and based on this, the drag coefficient optimization has been carried out by underbody flattening of the passenger vehicle.

2. Calculation method
2.1. Geometric Model
Due to the complex shape and structure of the vehicle, the body surface of the vehicle model and the interior of the engine compartment are properly simplified. The model should reflect the actual state as much as possible, especially the reaction on the aerodynamic performance of the vehicle when the underbody changed. Therefore, the model includes not only the integrity of the upper body geometry, but also completely retains the engine compartment and underbody components, which is shown in Figure 1.

Figure1. CFD calculation model of the vehicle body

IOP Conf. Series: Materials Science and Engineering **517** (2019) 012006 doi:10.1088/1757-899X/517/1/012006

The accuracy of CFD is closely related to the size of the calculation grid, but it is necessary to refine the grid of the body details in combination with hardware limitations. The main parameters of the calculation grid are shown in Table 1.

Table1. Description of the calculation grid

Type	Description
Surface unit	The basic size of the surface unit is 2-8mm, the size of encrypted parts (such as rear view mirror, rear wing, etc.) are 1-4mm, the size of the engine compartment interior parts are 4-16mm and the air intake grille size of the Engine compartment is 1-2mm.
Body unit	The minimum size of body unit encryption area is 16mm and the total grid number is up to 35 million.
Boundary layer unit	Body outer surface and accessories: 10 layers in total, the first layer thickness is 0.05mm and the total thickness is 6mm.

In the process of calculation, an area surrounding the vehicle is established to simulate the wind tunnel to monitor the state of the flow field, which is called the fluid calculation domain [5]. The calculation domain size is set to 66*12*10m, the front end of the passenger vehicle is 20m from the front end of the calculation domain, and the number of the body unit grid is about 35 million. The Realizable K-E turbulence model is used to set the boundary conditions, the Y+ value of the vehicle body surface is approximately 4-6. The front of the vehicle is set as the calculation domain entrance, the inlet turbulence ratio is 1%, and the speed is 33.33 m/s. The calculation domain outlet is the pressure outlet boundary, the ground is the non-slip wall boundary and the other calculation domain wall is the sliding wall boundary. The condenser and radiator use adopt porous media models and cooling fan is set not to rotate.

2.2. Calculation results

The pressure at each point in the calculation domain can be calculated, and the space corresponding to which called the pressure field. In order to describe the airflow visually, different colors which represent the pressure value are used to reflect the flow field pressure distribution. Figure 2 shows the pressure coefficient of vehicle body surface (Cp) and equipotential surface of the total pressure equaling to zero.

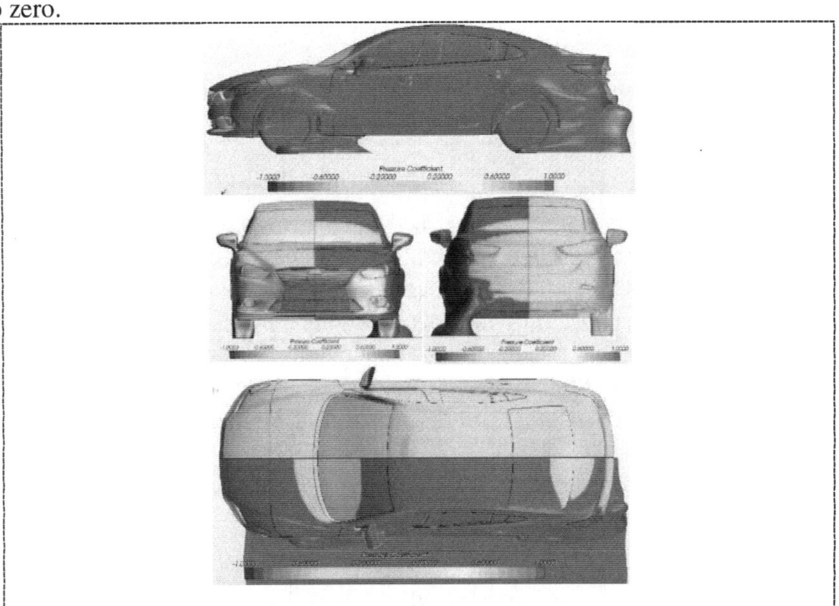

Figure2. Schematic of Cp and equipotential surface of the total pressure equaling to zero

The airflow is blocked when it encounters the front of the vehicle so that the speed of the airflow is greatly reduced and a positive pressure zone is produced, then the airflow flows in two directions. A part airflow flows to the top of the vehicle and when passing through the front windshield, the airflow once again is blocked and weakened to form a positive pressure zone. When reaching the highest point of the vehicle body, the larger airflow speed turns sharply to form a negative pressure zone. When arrived at the rear trunk lid, a sudden change in direction of airflow causes the airflow to separate and a small negative pressure zone is formed. Then the airflow accelerates tangentially along the upper surface of the trunk lid. The other part airflow flows from the bottom of the vehicle to the rear. When passing through the tire, a negative pressure zone is formed due to the effect of tires. When entering the bottom of the vehicle, a negative pressure zone is formed due to the narrow aisle and the fast speed of airflow.

The two parts of the airflow merge at the rear of the vehicle to form a vortex resulting in energy loss, which is one of important influence factors for aerodynamic drag. The calculation result predicts the flow field of the vehicle correctly and reflects the local airflow separation of the wheel cavity and the rearview mirror, etc.

2.3. Model reliability analysis and evaluation

In order to further verify the credibility of the geometric model and calculation method, a wind tunnel test is carried out in the aero-acoustic wind tunnel in Shanghai Ground Vehicle Wind Tunnel Center (shown in Figure 3).

Figure3. Aero-acoustic wind tunnel in Shanghai

The wind tunnel test is performed under the same conditions as CFD and the wind speed is set to 120km/h. By comparing the calculation result with the test result in three states (shown in Table 2), it is found that the test result is in good agreement with the CFD results, the relative deviation of the drag coefficient (Cd) is within 3%, and the CFD result is lower than the test result. The relative deviation of the lift coefficient (Cl) is slightly larger, but the maximum relative deviation is also about 5%. The overall evaluation shows that the calculation model and method are credible, and the model can be used to research the flattening underbody.

Table 2. The calculation result and the test result in three states

No.	Model for CFD and wind tunnel test	Cd			Cl		
		CFD	Test	Relative deviation	CFD	Test	Relative deviation
1#	Basic model (consistent with model 5 below)	0.315	0.323	2.5%	0.161	0.166	3.0%

2#	Plugging cooling grille based on 1#	0.295	0.299	1.3%	0.107	0.111	3.6%
3#	Removing the rearview mirror based on 2#	0.282	0.285	1.1%	0.159	0.151	-5.3%

3. Flattening underbody optimization

3.1. Model establishment

The unevenness of underbody hinders the airflow at the bottom of the vehicle. The underbody shape is more complex and the aerodynamic resistance is the greater, which will increase sharply with the increase of the vehicle speed. Thus it is useful to reduce the vehicle aerodynamic resistance, induced resistance and frictional resistance by optimizing the flattening underbody [6]. The underbody parts studied in this paper is shown in Figure 4.

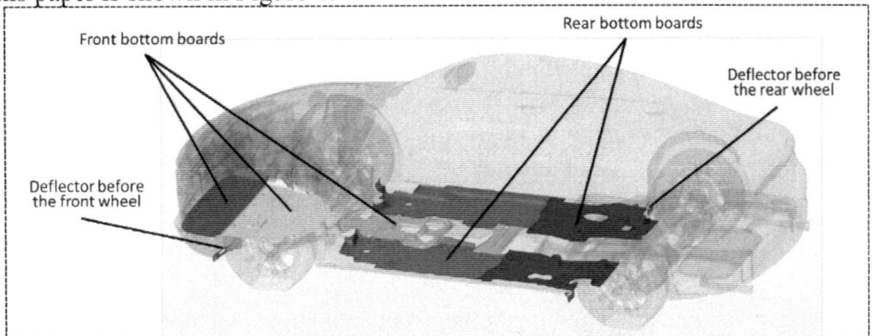

Figure4. The underbody parts studied

Five full-scale external flow field calculation models were established (Figure 5). The specific differences between models are: Model 1: Basic model. Model 2: Add four front bottom boards on Model 1. Front bottom boards is below the cooling module, engine compartment and the chassis central channel. Model 3: Add bottom boards below both sides of the central channel of the chassis, both sides of the tank and the rear suspension based on Model 2. Model 4: Add deflector before the front wheel based on Model 3. Model 5: Add deflector before the rear wheel based on Model 4.

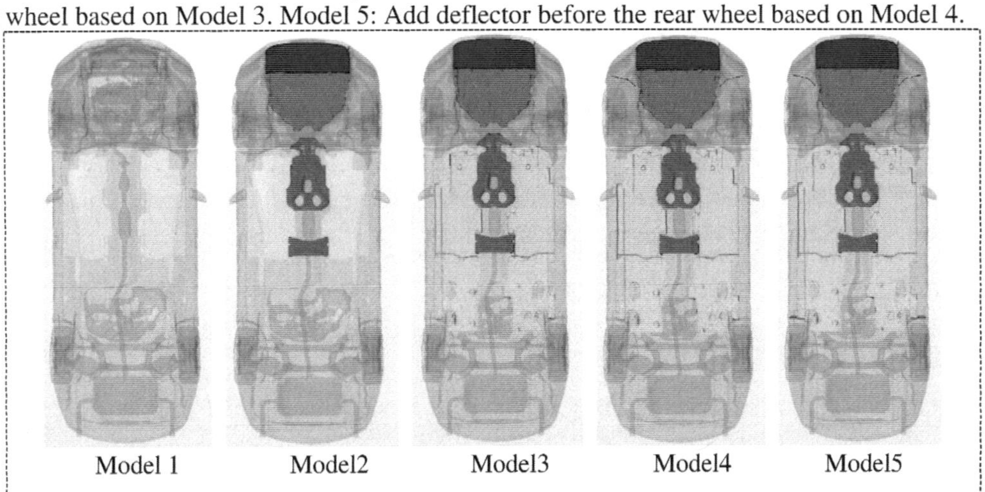

Figure5. The differences of 5 models

3.2. The cloud map of pressure coefficient

The cloud map of pressure coefficient for key parts of five models are obtained by calculations, as shown in Figure 6 and Figure 7.

Figure6. The cloud map of pressure coefficientof the bottom of Model 1,Model 2 and Model4 (from left to right)

The left picture of Figure 6 shows that: owing to the underbody of the model 1 not optimized, the pressure distribution below the engine and the cooling module are more complicated and not uniform. Air pressure increasing at the bottom of the vehicle when airflow passes through both sides of the central channel of the chassis, both sides of the tank and the rear suspension area. The pressure of the front and rear wheels is large which causes energy consumption of wheels. Contrasting the three pictures in Figure 6 shows that the pressure distribution of model 2 and model4 becomes even and the pressure decreases by adding bottom boards below the engine, chassis center, tank, rear suspension, etc.

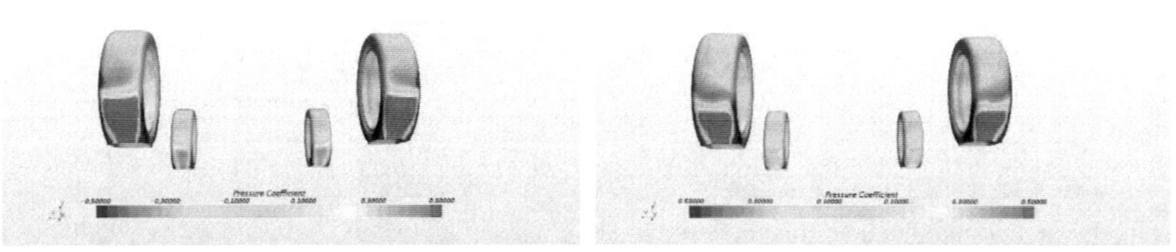

Figure7. The cloud map of pressure coefficient of front and rear wheels with and without deflectors installed before (from left to right)

The Figure 7 shows that the airflow pressure and the turbulence before front and rear wheels of model 4 and model 5 are reduced by adding deflectors, then reaching to reduce the resistance to wheels.

3.3. Contrast to Cd and Cl

The calculation results of the Cd and Cl of five models are shown in Table 3, which shows that the aerodynamic drag and lift of the vehicle have been variously varied by series of flattening optimization of underbody. The optimal optimization has an effect of 18% on drag and 31% on lift.

Table 3. The results of calculating the Cd and Cl of five models

Model	Cd	Relative value of Cd	Cl	Relative value of Cl
1	0.387	1.000	0.199	1.000
2	0.350	0.904	0.149	0.749
3	0.335	0.868	0.136	0.683
4	0.317	0.819	0.168	0.844
5	0.315	0.814	0.161	0.809

Analyzing the relative values of Cd and Cl in Figure 8 shows that the effect on the aerodynamics of the vehicle of different components. Both bottom boards and deflectors before wheels have the effect of reducing the aerodynamic drag, and the most obvious effect comes from the rear four-piece boards and the deflector before front wheels. All bottom boards are beneficial to reduce aerodynamic lift, which is mainly due to that the flattening underbody lets the airflow accelerated and the pressure reduced at the bottom of the body. It is noteworthy that the deflector before the front wheel increases the aerodynamic lift of the vehicle greatly, but the deflector before rear wheels slightly reduces the aerodynamic lift of the vehicle.

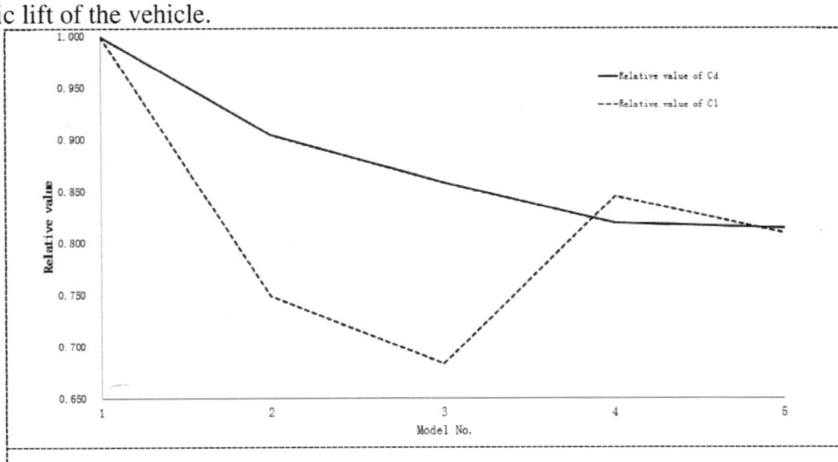

Figure8. Relative values of Cd and Cl

4. Conclusions

Based on the effectiveness of CFD on the aerodynamics calculation, the drag and lift coefficients Cd and Cl of the vehicle were predicted with error not more than 5%. With underbody flattening, the drag coefficient Cd was reduced 18% and the lift coefficient Cl was reduced 31% maximum in five models with different flattening degree, which shows the significance on the aerodynamic performance of the vehicle. It is noteworthy that the deflector before wheels will reduce the aerodynamic drag of the vehicle, but which has an adverse effect on the lift.

5. References

[1] GU Zhengqi. Automobile Aerodynamics, M. People's Communication Press, 2005.

[2] Sun S, Chang Y, Fu Q, et al. Aerodynamic Shape Optimization of an SUV in early Development Stage using a Response Surface Method, J. SAE Technical Paper 2014-01-2445.

[3] Thomas Blacha, Moni Islam. The Aerodynamic Development of the new Audi Q5, J. SAE Int. J. Passeng. Cars-Mech.syst.10(2):2017, doi: 10.4271/2017-01-1522.

[4] ZHANG Yangjun, LV Zhenhua, XU Shian, et al. Advances in Numerical Simulation of Automobile Aerodynamics, J. Automotive Engineering, 2001, 23(2), pp.82-91.

[5] ZHANG Jianli, ZHANG Guozhong, Application of CFD in Automotive Aerodynamics Research, J. Journal of Shenyang University (Natural Science), 2006, 18(4), pp.15-18.

[6] Andreas Kremheller. The Aerodynamics Development of the New Nissan Qashqai, J. SAE Technical Paper. 2014.

6. Acknowledgments

This work is supported by the Key Special Technology Innovation Fund of Chongqing Science and Technology Commission (cstc2015zdcy-ztzx60011) and the Opening Fund of State Key Laboratory of Vehicle NVH and Safety Technology (NVHSKL-201402).

Dynamics Analysis of CBR Reducer with Tooth Modification

XiaoXiao Sun[1], Liang Han[1, a]

[1]School of Mechanical Engineering, Southeast University, Nanjing, P.R. China

[a]melhan@seu.edu.cn

Abstract. China Bearing Reducer (CBR) is a one stage new type cycloid drive reducer which has large transmission ratio, high payload, high torsional stiffness, high tilting stiffness, compact size. In this paper, the cycloid gear with tooth modification of CBR reducer was investigated to analyse the contact force and stress. Firstly, the mechanical design of CBR reducer was introduced. Then, the profile equation of cycloid gear was derived by inverted gear train method. And the modified profile equation was derived by adding a small increment into the profile equation. Secondly, the dynamic model was built by multi-dynamics theory. Thirdly, the number of simultaneous contact teeth, contact force, stress with respect to modification value were investigated by dynamics software. The results show that the larger is the clearance between the cycloid teeth and pin, the larger are the contact force and stress.

1 Introduction

The cycloid drive has the advantages of large reducer ratio, high torque, high precision, high stiffness and compact structure. It has been widely used in precision reducer, such as RV reducer. In this paper, the CBR reducer is introduced and the dynamics characteristics are analysed.

Many researches had been done by scholars on cycloid-pin gear reducer. Yang and Blanche [1,2] discussed the formulas of cycloid drive and investigated the effect of machining tolerances on backlash and torque ripple. Then they presented an analytical and computer-aided analysis and synthesis of cycloid drives. Teruaki Hidaka et al. [3] proposed a method by using equivalent dynamic model with equivalent error to analyze rotational transmission error. Subsequently, the effects of machining and assembly errors of elements on the rotational transmission error were investigated [4]. And the mutual effects of error of the elements on rotational transmission error were analyzed [5]. Xin Li et al. [6] proposed a double crank ring-plate-type cycloid drive which does not need output unit and places no limit on the size of the tumbler bearing, and working principles, advantages and design issues were discussed. Ta-Shi Lai [7] presented a cycloid drive with roller drive, and the equation of geometric design and meshing is derived. Joong-Ho Shin and Soon-Man Kwo [8] proposed an approach to derive cycloid profile by means of the principle of the instant velocity center. The epicycloid and hypocycloid were both discussed. Anh Duc Pham et al. [9] presented an efficient FE analysis procedure for the hysteresis characteristics of a cycloid reducer using a nonlinear spring with a dead zone. Zhong-Yi Ren et al. [10] proposed a new tooth modification method of cycloid disc by adjusting the position of 5 key points.

The aim of the paper is to investigate the contact force of cycloidal gear reducer. It will be expected to analyze the load sharing among the multiple contact tooth pairs, the distribution and variation of the contact stress. Three different modification values for force analysis of the CBR reducer are presented in the paper.

IOP Conf. Series: Materials Science and Engineering **517** (2019) 012007 doi:10.1088/1757-899X/517/1/012007

2 Design and profile equation of cycloid gear

2.1 Mechanical design

CBR reducer is a new type of cycloid gear reducer, which adopts a symmetrical transmission structure. As shown in Fig.1, it is mainly composed of bearing end cover, outputting flange, cross roller bearing, crank bearing, disc connector, turning bearing, crank shaft, pins, cycloid gear, case, outputting bearing and input support. Compared with RV reducer, it has the advantages of simple structure, small number of parts, simple assembly process and low manufacturing cost.

Because of its compact structure, it can achieve the installation dimension of the harmonic drive reducer. So, it can be used in robot joints with space restricted instead of harmonic reducer, and its payload, torsional stiffness and instantaneous impact resistance are several times higher than that of the harmonic drive reducer.

At the same time, in order to improve the transmission accuracy, we designed a disc connector between the output flange and the cycloid gear. Through the disc connector, the differential motion produced by the cycloid gear and the pins is transferred to the output flange. As shown in Fig. 2, there are two pairs of rounded rectangular holes on the disc connector, which are connected with the boss of the cycloid gear and the boss of the output flange respectively. In order to improve the transmission efficiency, rollers are placed between rounded rectangular hole and boss of cycloid gear and output flange to avoid sliding friction.

The parameters of CBR25 are shown in Table 1, we will analyse this type in later section.

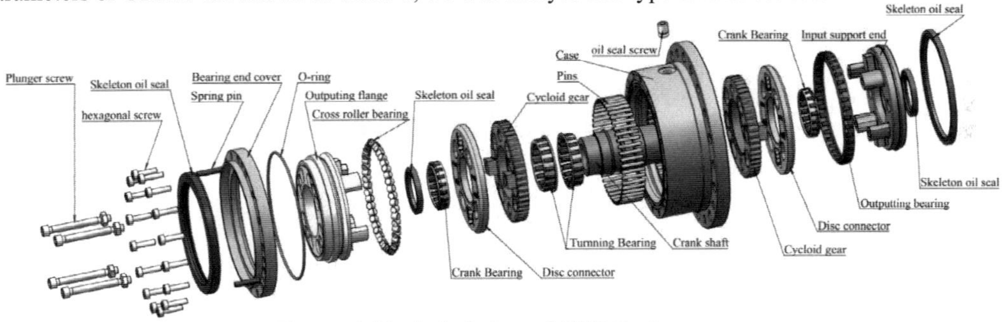

Figure 1. Exploded view of CBR Reducer.

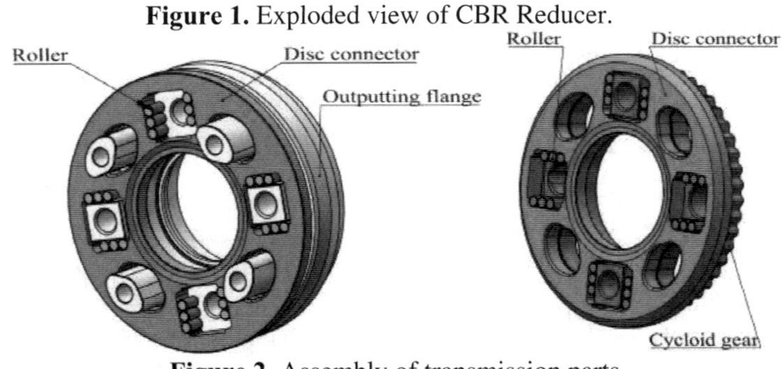

Figure 2. Assembly of transmission parts.

The parameters of CBR25 are shown in Table 1, we will analyse this type in later section.

Table 1. Parameters of CBR25 reducer.

Radius of pin position r_p	29.6
Pin radius r_{rp}	0.975
Teeth number of Pins z_p	50
Teeth number cycloid gear z_c	49
Transmission ratio i	49
Eccentricity a	0.462
Short width coefficient K_1	0.7804
Tooth width w_d	6.3

2.2 Tooth profile with modification

The analysis method of inverted gear train coordinate systems(IGTCS) is used to establish standard profile equation. As shown in Fig.3, an overall fixed coordinate system $S_f\,(X_f{-}O_f{-}Y_f)$ and a moving coordinate system

$S_p(X_p-O_p-Y_p)$ fixed to the pin gear are established in the center of the pin gear. And a moving coordinate system $S_c(X_c-O_c-Y_c)$ is established in the centre of the cycloid gear. In the initial position, the Y_p axis coincides with the Y_f axis, and the Y_c axis is parallel to the Y_f axis. In the IGTS, the cycloid gear rotates φ_c at the angular speed w_c counter clockwise around the point O_c (that is, S_c rotates φ_c around O_c). According to the relative motion relation, the pin gear will rotate φp at the angular speed w_p counter clockwise around the point O_p (that is, S_p rotates φ_p around O_p). So, we can get

$$i^H = w_c/w_p = \varphi_c/\varphi_p = z_p/z_c \qquad (1)$$

Where i^H is the transmission ratio of the cycloid gear and the pin gear. z_c is the number of teeth of the cycloidal gear and z_p is the number of teeth of pins.

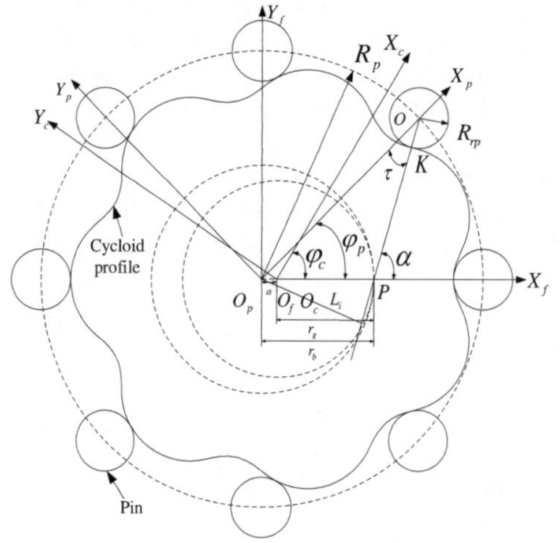

Figure 3. Standard tooth profle of cycloid disc.

Based on the theory of gear meshing and differential geometry, the standard cycloid profile in S_c can be expressed as

$$\begin{cases} x_c = \left(R_p - R_{rp}S_r^{-\frac{1}{2}}\right)\cos[(1-i^H)\varphi_p] - \left(a - K_1R_{rp}S_r^{-\frac{1}{2}}\right)\cos(i^H\varphi_p) \\ y_c = \left(R_p - R_{rp}S_r^{-\frac{1}{2}}\right)\sin[(1-i^H)\varphi_p] + \left(a - K_1R_{rp}S_r^{-\frac{1}{2}}\right)\cos(i^H\varphi_p) \end{cases} \qquad (2)$$

Where, R_p is the radius of pin gear and R_{rp} is the radius of the pin, a is the eccentricity of the input crank shaft, K_1 is short width coefficient, $K_1 = az_p/R_p$. $S_r = 1 + K_1^2 - 2K_1\cos\varphi_p$.

To compensate for the errors caused by assembly and manufacturing, it is necessary to modify the cycloidal gear profile. Three ways of modification are commonly used: isometric modification (modifying the pin radius R_{rp}), offset modification (modifying the pin gear radius R_p) and angle rotation modification (modifying the rotation angle φ_c of cycloid gear). It is often the compound use of these three methods, especially the compound uses of offset modification and isometric modification. The cycloidal gear profile with modifications can be obtained by adding a small increment of the pin center position ΔR_p to R_p, a small increment of the pin radius Δr_{rp} to r_{rp} and a small increment of the cycloid disc rotational angle δ to φ_c in (2). The equation for the modifed profle can be written as (3).

$$\begin{cases} x'_c = \left((R_p+\Delta R_p) - (R_{rp}+\Delta R_{rp})S_r^{-\frac{1}{2}}\right)\cos[(1-i^H)\varphi_p - \delta] - \\ \frac{a}{R_p+\Delta R_p}\left((R_p+\Delta R_p) - z_p\left((R_{rp}+\Delta R_{rp})S_r^{-\frac{1}{2}}\right)\right)\cos(i^H\varphi_p+\delta) \\ y'_c = \left((R_p+\Delta R_p) - (R_{rp}+\Delta R_{rp})S_r^{-\frac{1}{2}}\right)\sin[(1-i^H)\varphi_p - \delta] + \\ \frac{a}{R_p+\Delta R_p}\left((R_p+\Delta R_p) - z_p\left((R_{rp}+\Delta R_{rp})S_r^{-\frac{1}{2}}\right)\right)\sin(i^H\varphi_p+\delta) \end{cases} \qquad (3)$$

3 Dynamic modelling

IOP Conf. Series: Materials Science and Engineering **517** (2019) 012007 doi:10.1088/1757-899X/517/1/012007

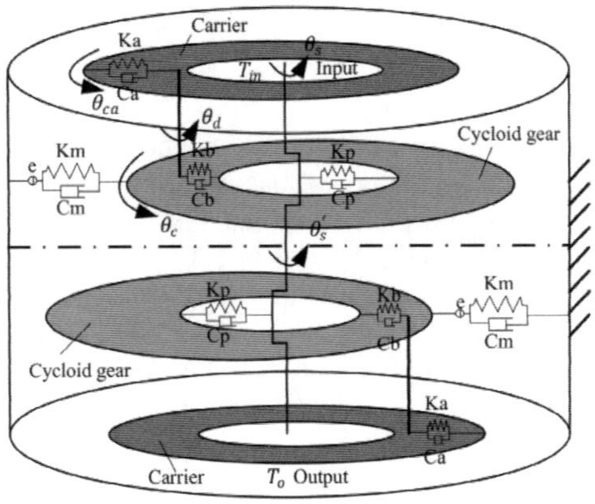

Figure 4. Dynamic model of CBR reducer.

Tooth modification will affect the dynamic performance of cycloidal speed reducer. In this paper, neglecting the stiffness of the bearing, an eight degrees of freedom (DOF) discrete dynamic model of CBR reducer is established as shown in Fig. 4. As illustrated in Fig. 4, the structure of CBR reducer is symmetrical, so we can build the dynamic model with half of it, just four DOFs. Each component is represented by a rigid disc of radius R and polar mass moment of inertia *J*. The cycloid disc mesh is represented by two sets of periodically time–varying mesh spring–damping elements with $K_m(t)$ and $C_m(t)$. The crank shaft and cycloid disc mesh is represented by two sets of spring damping elements with K_p and C_p. The cycloid disc and disc connector mesh is represented by two sets of spring damping elements with K_b and C_b. The contact of disc connector and carrier is represented by two sets of spring damping elements with K_a and C_a.

With the positive directions of the alternating rotational displacements θ_s, θ_c, θ_d, θ_{ca} and the constant external torques T_{in} and T_o are defned in Fig. 4, the equations of motion of the cycloidal speed reducer are written as:

$$\begin{cases} J_s\ddot{\theta}_s + [K_p(\theta_s - \theta'_s) + C_p(\dot{\theta}_s - \dot{\theta}'_s)] \cdot a = T_{in} \\ J_c\ddot{\theta}_c + \sum_{i=1}^{n}[K_{m_i}(\theta_{c_i} - \theta_r) + C_{m_i}(\dot{\theta}_{c_i} - \dot{\theta}_r)] \cdot L_i = \\ \quad [K_b(\theta_{c_i} - \theta_d) + C_b(\dot{\theta}_{c_i} - \dot{\theta}_d)] \cdot R_1 + \\ \quad [K_p(\theta_c - \theta'_c) + C_p(\dot{\theta}_c - \dot{\theta}'_c)] \cdot R_2 \\ J_d\ddot{\theta}_d + [K_p(\theta_d - \theta_o) + C_p(\dot{\theta}_d - \dot{\theta}_o)] \cdot R_1 = \\ \quad [K_a(\theta_o - \theta'_o) + C_a(\dot{\theta}_o - \dot{\theta}'_o)] \cdot R_3 \\ J_{ca}\ddot{\theta}_{ca} + [K_a(\theta_o - \theta'_o) + C_a(\dot{\theta}_o - \dot{\theta}'_o)] \cdot R_3 = T_o \end{cases} \quad (4)$$

where, $\theta_s - \theta'_s$ is the torsional angle of crank shaft, θ_c - θ'_c is the relative displacement of cycloid disc and crank shaft, and θ_o - θ'_o is the torsional angle of carrier. Subscript *i* is represents the *i*-th cycloid disc meshing tooth pair. *n* is the total number of cycloid disc meshing tooth pair. L_i is the arm of cycloidal tooth force as shown in Fig. 3.

4 Results and discussions

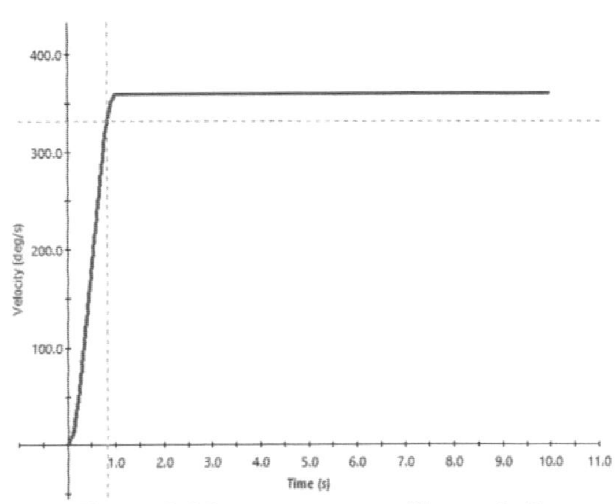

Figure 5. Movement curve of input shaft.

4.1 Conditions of the Dynamic Simulation

In order to analyze the influence of the modifcation on the dynamic performance of the CBR reducer, we use the modules of motion and simulation of SolidWorks to analyze the contact force and stress of the CBR reducer in this paper. To study the influence of the clearance between the teeth on the performance, a single isometric modifcation method is adopted because the normal gap between the teeth generated by isometric modifcation is equal. The number of simultaneous contact teeth, contact force and stress are analyzed under modifcation values of 10 μm, 20 μm and 30 μm, respectively. The rotational speed of input shaft is 60 rpm (360 deg/s). The given movement curve for the input crank shaft is as shown in Fig. 5. The acceleration phase occurs between 0 and 1 s, and the velocity is constant after 1 s. The gravitational acceleration is 9.8 *m/s2*. The parts are made of alloy steel, and the relevant mechanical properties are as listed in Table 2.

Table 2. Mechanical properties of material.

Mechanical properties	Values (Unit: $/mm^2$)
Elastic modulus	210,000
Shear modulus	79,000
Tensile strength	210,000
Yield strength	79,000

4.2 Contact force of each tooth

The forces acting on each tooth of the cycloidal gear are different when input crank shaft rotates at different degree. In order to express the force size of each tooth at different degree vividly, the force vector is used to express the direction and magnitude of each tooth. As shown in Fig.6, the force is expressed when the input crank shaft rotate at 0 deg,72 deg, 144 deg, 216 deg, 288 deg and 360 deg with the load torque of 100 N(in SolidWorks motion software). It can be seen from the diagram that the force acting lines of each tooth are all along the common normal line of the meshing line, pointing to the same point which is the internal tangential point of pitch circle the pins and the cycloid disc.

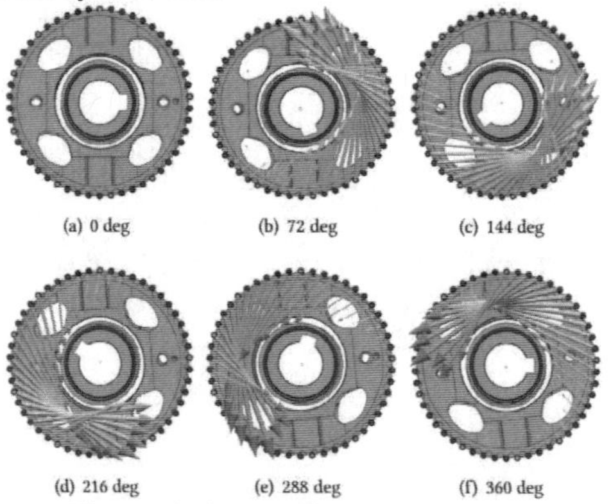

Figure 6. Contact force distribution.

4.3 Number of simultaneous contact teeth influenced by tooth modifcation

All the cycloid gear teeth are in contact with the pin teeth in the theoretical profle without modifcation. After the modifcation, the cycloid profle and the pins are not satisfed with the conjugate condition. Only one cycloid tooth is contact with pin after a small angle of rotation of cycloid gear. With the increase of load torque, the contact deformation between cycloid teeth and pin teeth is increased, and the number of contact teeth increases gradually. The number of contact teeth corresponding to different modifcation values are shown in the Fig.7. It can be seen from the diagram that under the same load torque, the larger the modifcation value is, the less the contact tooth number is. This is because the increase of the modifcation value causes the increase of the gap between the teeth, and the same size of the load torque cannot make the cycloid tooth produce large contact deformation to offset the gap.

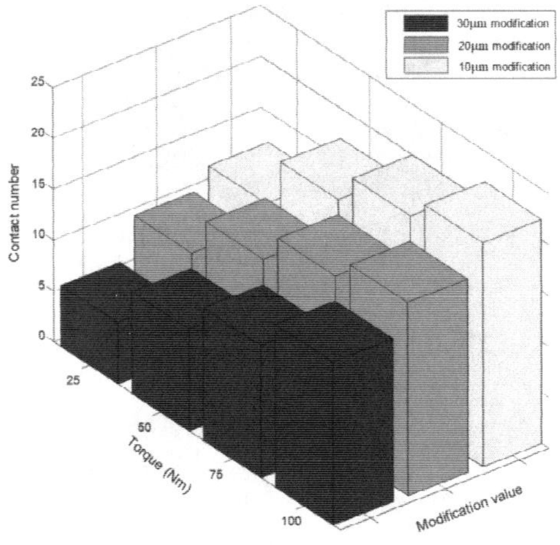

Figure 7. Contact number with respect to output torque.

4.4 Contact force influenced by tooth modification

Not considering the deformation, only one pair tooth contact with tooth modifcation, so the cycloid gear will produce enormous contact force. The actual situation is that the cycloid gear and the pins will have deformation. According to the size of the deformation and the Hertz contact theory, there will be different contact forces between the cycloidal teeth and the pin teeth. The frst contact tooth has the maximum force, the last contact tooth has the minimum force, and the uncontacted teeth have no forces. The forces of different modifcation are shown in Fig. 8 with the load torque of 100 N.

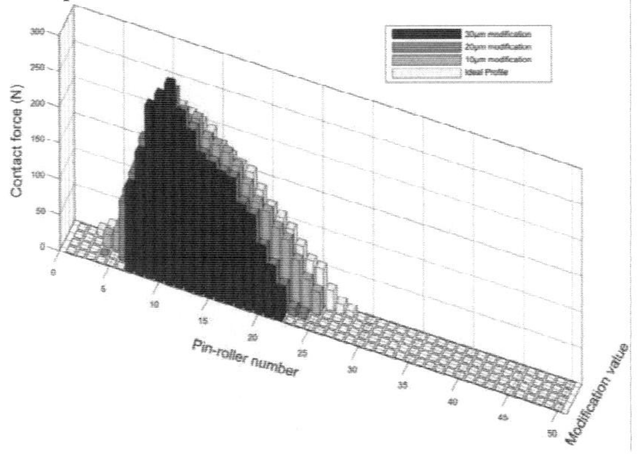

Figure 8. Contact forces distribution of cycloidal tooth.

4.5 Stress of cycloid gear influenced by tooth modification

The contact stress between teeth can be calculated by fnite element method(FEM). SolidWorks simulation is the FEM module, and it can connect seamlessly with the motion module. The constraints and the force of the analysis result of the motion module can be imported into the simulation module for transient dynamic analysis. Here, the contact stress of cycloid gear is the focus. We only mesh the cycloid gear and calculate its stress. The meshing structure of cycloid gear is shown in Fig.9, where the number of elements are 32007 and the number of nodes are 65321. In Fig.10, the stress cloud of cycloid gear is shown in the same load torque with different modifcation values. We can see that the stress with no modifcation is smallest, the stress with 30 μm modifcation is largest.

Figure 9. Mesh of cycloidal gear.

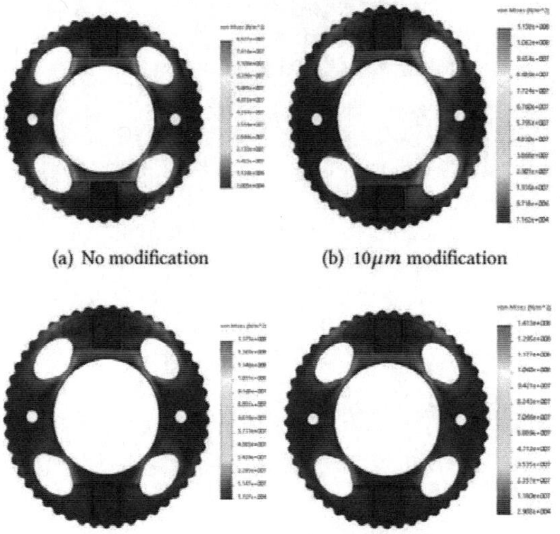

(a) No modification (b) $10\mu m$ modification

(c) $20\mu m$ modification (d) $30\mu m$ modification

Figure 10. Stress cloud.

5 Conclusion

This article presents the design and dynamics simulation of CBR reducer. CBR is a one-stage reducer with a compact structure and a wide range of installation size. The modifed profle equation of cycloid gear was derived and the multibody dynamic equation was built. Different modifcation values were used to built CAD model for dynamics analysis to investigate the number of simultaneous contact teeth, contact force, contact stress with respect to modifcation value. The simulation results showed that the large gap between the teeth will cause large force and stress. So a suitable gap can better balance the contact force and thermal expansion of CBR reducer.

Acknowledgment

This project is supported by Science and Technology Program of Jiangsu Province (Grant No. SBE2015000030), the Fundamental Research Funds for the Central Universities and the Research Innovation Program for College Graduates of Jiangsu Province (Grant No. KYLX16_0187). These supports are gracefully acknowledged.

References

[1] J. G. Blanche and D. C. H. Yang, "Cycloid drives with machining tolerances," *J. Mech. Transm. Autom. Des.*, vol. **111**, no. 3, pp. 337–344, (1989).
[2] D. C. H. Yang and J. G. Blanche, "Design and application guidelines for cycloid drives with machining tolerances," *Mech. Mach. Theory*, vol. **25**, no. 5, pp. 487–501, (1990).
[3] T. Hidaka, H. Y. Wang, T. Ishida, K. Matsumoto, and M. Hashimoto, "Rotational transmission error of k-h-v planetary gears with cycloid gear : 1st report, analytical method of the rotational transmission error," *Trans. Jpn. Soc. Mech. Eng. Ser.C*, vol. **60**, no. 570, pp. 645–653, (1994).

[4] T. Ishida, H. Wang, T. Hidaka, and M. Hashimoto, "Rotational transmission error of k-h-v-type planetary gears with cycloid gears. 2nd report, effects of manufacturing and assembly errors on rotational transmission error." *Trans. Jpn. Soc. Mech. Eng. Ser. C*, vol. **60**, no. 578, pp. 3510–3517, (1994).

[5] H. Wang, T. Ishida, T. Hidaka, and M. Hashimoto, "Rotational transmission error of k-h-v-type planetary gears with cycloid gear : 3rd report, mutual effects of errors of the elements on the rotational transmission error," *Trans. Jpn. Soc. Mech. Eng. Ser. C*, vol. **60**, no. 578, pp. 3518–3525, (1994).

[6] X. Li, W. He, L. Li, and L. C. Schmidt, "A new cycloid drive with high-load capacity and high efciency," *J. Mech. Des.*, vol. **126**, no. 4, pp. 683–686, (2004).

[7] T. S. Lai, "Geometric design of roller drives with cylindrical meshing elements," *Mech. Mach. Theory*, vol. **40**, no. 1, pp. 55–67, (2005).

[8] J. H. Shin and S. M. Kwon, "On the lobe profle design in a cycloid reducer using instant velocity center," *Mech. Mach. Theory*, vol. **41**, no. 5, pp. 596–616, (2006).

[9] A. D. Pham, T. L. Tran, and H. J. Ahn, "Hysteresis curve analysis of a cycloid reducer using non-linear spring with a dead zone," *Int. J. Precis. Eng. Manuf.*, vol. **18**, no. 3, pp. 375–380, (2017).

[10] Z. Y. Ren, S. M. Mao, W. C. Guo, and Z. Guo, "Tooth modifcation and dynamic performance of the cycloidal drive," *Mech. Syst. Signal Process.*, vol. **85**, pp. 857–866, (2017).

[11] K. S. Lin, K. Y. Chan, and J. J. Lee, "Kinematic error analysis and tolerance allocation of cycloidal gear reducers," *Mech. Mach. Theory*, vol. **124**, pp. 73–91, (2018).

IOP Conf. Series: Materials Science and Engineering **517** (2019) 012008 doi:10.1088/1757-899X/517/1/012008

A Novel Distributed and Compound Control of an Adaptive Neural Network and PD for Manipulators

Pu Zhao[1], **Yunfei Zhou**[1]
[1] School of Mechanical Science and Engineering, Huazhong University of Science and Technology, 1037 Luoyu Road, Wuhan 430074, China

Abstract. In this report, a new compound control scheme is proposed for manipulators based on a proportional-differential (PD) controller and a back-propagation neural network. By modifying the dynamic model of manipulators, a second-order function of tracking errors is obtained and approximated by the neural network. A PD-type reference model is introduced to update the weights of the neural network. By using additional PD controllers, rapid response of the control scheme can be realized. Simulation results highlight performance of the controller to compensate the nonlinear terms and its simpleness in the parameter tuning process. To be concluded, the controller is suitable for distributed control and can be used as supplementary of traditional PD controllers.

1. Introduction

Artificial neural networks, as a kind of tools of computational intelligence, have been credited in various applications as powerful tools capable of providing robust controllers for mathematically ill-defined systems that may be subjected to structured and unstructured uncertainties[1]. The universal approximation theorem has been the main driving force behind the increasing popularity of such methods, as it shows that they are theoretically capable of uniformly approximating an continuous real function to any degree of accuracy. This has led to recent advances in the area of intelligent control[2]. Various neural networks have been applied in the control of manipulators, which have led a satisfactory performance[3]. Ponce proposed the design and implementation of a new neural network based on a trigonometric series expansion and applied it to the control of a nonlinear system[4]. Hui suggested an observer-based adaptive controller using time-delay neuro-fuzzy network for manipulators with flexible joints[5]. Chaoui designed an adaptive controller based on two backpropagation neural networks for manipulators with Coulomb friction model and flexible joints[6]. Recently, adaptive fuzzy neural networks are used by Chen to approximate a nonlinear stochastic system with unknown functions [7]. In [8], neural networks are used to approximate nonlinear models of friction and backlash hysteresis. In [9], a trajectory tracking controller, including a neural adaptive compensator, is proposed by Rossomando for a unicycle link mobile robot. Le proposed an adaptive controller with orthogonal neural network for manipulators [10]. In general, the presence of high, particularly unstructured, nonlinearities, such as in the form of Coulomb friction on a manipulator makes it unrealizable to solve the inverse dynamics of the system. Hence, the controllers designed by the researchers adopted desired position as the input of the neural networks, such as Chaoui and Ye [6,11]. However, this leads to unachievable of distributed control of manipulators. In this report, by modifying the dynamic model of manipulators, a second-order function of tracking errors is obtained and approximated by neural networks. Different with existing methods, this method utilize the tracking errors rather than desired trajectory of the manipulators and thus can be seemed as a distributed controller. As shown in the simulation results, tracking errors and response speed can be tuned easily by using the same tuning method of PD controllers.

2. Dynamics of manipulators

The dynamics of a serial n-link rigid manipulator can be written as

$$M(q)\ddot{q}+C(q,\dot{q})\dot{q}+G(q)+F(\dot{q})+T_d(q,\dot{q},\ddot{q})=\tau, \tag{1}$$

where q, $\dot{q}$ and $\ddot{q}$ are $n\times1$ vectors of joint angle, angle velocity and angle acceleration, respectively; τ is an $n\times1$ vector of input torques; $M(q)$ is an $n\times n$ symmetric positive-definite inertia matrix; $C(q,\dot{q})$ is an $n\times n$ matrix of centripetal and Coriolis torques; $G(q)$ is an $n\times1$ vector of gravitational torques obtained as the gradient of the robot potential energy due to gravity; $F(\dot{q})$ is an $n\times1$ vector for the friction torques; $T_d(q,\dot{q},\ddot{q})$ is an n×1 vector of disturbance torques.

The dynamics of a robotic manipulator are characterized by the following property:

Property 1: The inertia matrix $M(q)$, Coriolis matrix $C(q,\dot{q})$ and gravity vector $G(q)$ are upper and lower bounded, i.e., there exist scalars $\alpha_1(q)$, $\alpha_2(q)$, $\alpha_3(q)$ and $\alpha_4(q)$ such that $\alpha_1(q)I\le M(q)\le\alpha_2(q)I$, $\|C(q,\dot{q})\dot{q}\|\le\alpha_3(q)\|\dot{q}\|^2$ and $\|G(q)\|\le\alpha_4(q)$, where I is the identity matrix.

Before we proceed further, we introduce the following realistic assumption:

Assumption 1: The norm of the unknown disturbance T_d is upper bounded by a scalar α_5, i.e., $\|T_d\|\le\alpha_5$.

3. Control Strategy

The control strategy is based on the design of a reference model based neural network controller and a PD controller. After analysing the characteristics of the two different controllers and dynamics of manipulators, a new compound control scheme is proposed.

3.1. Neural network and reference model

NN-based controllers can overcome the difficult control problems in nonlinear control systems due to the nonlinear approximation capabilities of NNs. Furthermore, NN controllers have been successfully deal with robot control problems mentioned by many researchers. In this report, the neural network proposed in [12] is selected as the network controller for its Lyapunov stability. The relationship between the inputs and outputs of a neural network can be described by

$$v_j^{(l)}(k)=\sum_{i=1}^{m}\omega_{ji}^{(l)}(k)x_i^{(l)}(k), O_j^{(l)}(k)=\varphi_j^{(l)}\left(v_j^{(l)}(k)\right), \tag{2}$$

where m and n being the number of input and hidden nodes, respectively; φ is the sigmoid activation function. The signal $O(k)$ represents the output of the neural network. The new weight matrices are computed by the rules modified from

$$\omega_{1j}^{(2)}(k)=\frac{d(k-1)+\eta e(k-1)}{O_j^{(1)}(k-1)+\delta_1}, \omega_{ji}^{(1)}(k)=\frac{1}{mx_i(k)+\delta_2}\varphi^{-1}\left(\frac{d(k-1)+\eta e(k-1)}{O_j^{(1)}(k-1)+\delta_3}\right), \tag{3}$$

where η is the network's learning rate. δ_1, δ_2 and δ_3 are small values to avoid numerical singularity problems. Based on the results presented in [12], δ_1, δ_2 and δ_3 don't affect the convergence of the neural network when selected small enough. $d(k-1)$ is the expected output value of the neural network for the instant $(k-1)$, while $e(k-1)$ is the output error of the network and satisfies $d(k-1)=y(k-1)-e(k-1)$.

Theorem 1: The learning algorithm (3) with a learning rate η satisfying $0<\eta<1$ guarantees the bounded stability of the neural network.

Proof: Consider the following candidate Lyapunov function:

$$V(k)=\frac{1}{\eta}e^2(k),\ \ \Delta V(k)=\frac{1}{\eta}e^2(k)-\frac{1}{\eta}e^2(k-1). \tag{4}$$

Based on the results presented in [6], it's easy to obtain the following equations

$$\begin{aligned}\Delta V(k)&=\frac{1}{\eta}\left(\eta e(k-1)+d(k-1)-d(k)\right)-\frac{1}{\eta}e^2(k-1)\\ &\le \eta e^2(k-1)-\frac{1}{\eta}e^2(k-1)+\frac{4}{\eta}\left(d(k-1)-d(k)\right)^2\le\eta\left(1-\frac{1}{\eta^2}\right)e^2(k-1)+\frac{4}{\eta}\Omega^2,\end{aligned} \tag{5}$$

where Ω denotes the difference between the desired output on instant k and k-1. Based on the Property 1 and Assumption 1 mentioned in the Section 2, Ω is bounded by a specific value. By using (4), the output error $e(k)$ satisfies the following equation

$$e^2(k)\le\eta^2e^2(k-1)+4\Omega^2\le\eta^2\left(\eta^2e^2(k-2)+4\Omega^2\right)+4\Omega^2\le\eta^{2k}e^2(0)+\frac{1-\eta^{2k}}{1-\eta^2}4\Omega^2. \tag{6}$$

As a result, the output error of the neural network converges exponentially to a bounded value when $0<\eta<1$.

Remark: It should be noted that the region becomes smaller for slowly time-varying systems or for fast neural network sampling rates. Besides, this modified proof process also shows that the learning rate affects the region and convergence speed of the NN.

The reference model is designed to calculate the output error of the neural network. Let $\Delta q = q_d - q$ denote the links' position with q_d being the desired motor position vector. A PD-type reference model is selected to calculate the output error and presented as $e(k)=s=K_{ps}\Delta q+K_{ds}\Delta\dot{q}$. K_{ps} and K_{pd} are proportional gain and derivative gain of the reference model. Since the reference model has the same structure with PD controller, the tuning methods of the parameters are the same.

3.2. Compound control scheme

Robotic manipulator dynamics with multi-joints are highly nonlinear, highly coupling and form an uncertain model system. Although the PID controller has a simple form and robustness in broad operating conditions, single PID controllers cannot reach high control precision in uncertain nonlinear control systems. As a result, a compound control method based on NN and PD is introduced for the position-tracking control of manipulators. To decide the number of input variables of the neural network, modifications of (1) are presented as follows:

$$\begin{aligned}\tau&=M(q)\ddot{q}+C(q,\dot{q})\dot{q}+G(q)+F(\dot{q})+T_d(q,\dot{q},\ddot{q})\\ &=f(q,\dot{q},\ddot{q})=\mathrm{g}(\Delta q,\Delta\dot{q},\Delta\ddot{q}),\end{aligned} \tag{7}$$

where f and g are unknown nonlinear functions. Based on(7), desired output of the neural network can be treated as a second-order nonlinear filter. Besides, to reduce the input of the networks, the coupling factors for each link are also approximated by the networks. As a result, the input variables for joint i is $\left[e_i(k),e_i(k-1),e_i(k-2)\right]$.

Figure 1 shows a block diagram representation of the control system of each joint. In Figure 1, q_d and q represent the desired and actual position of the corresponding manipulator joint, respectively. u_{PD} and u_{NN} are the output of PD controller and NN controller. In the control process, the PD controllers reach system stability and help achieve rapid response. Then, the NN controllers can learn in an on-line continuous learning manner to realize the forward compensation control for the dynamic model of the manipulator.

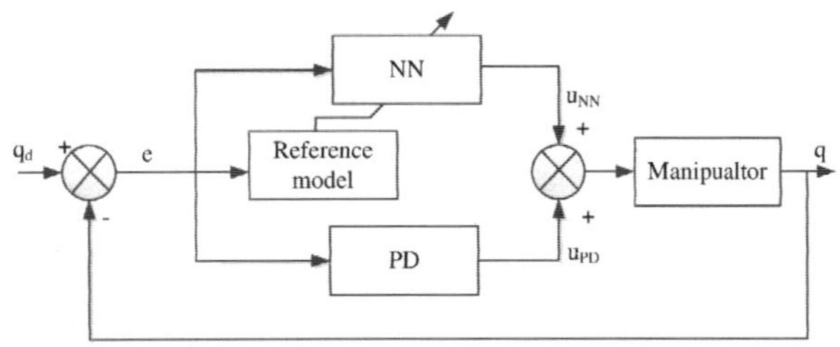

Figure 1. Proposed control scheme.

4. Simulation results and discussions

In order to show the effectiveness of the proposed control system, simulation experiments are carried out. In this section, we consider the first three links of Puma P560. The dynamics of the manipulator can be obtained by Robotics Toolbox for Matlab. The compound controller of the NN and PD and the single PD controller for the three-link model are given to compare their control performance.

In the simulation experiments, sampling time is selected as 1ms. As the for the friction, the Coulomb parameter and dynamic friction coefficient are selected as 0.3 and 0.2, respectively. In order to consider the input constraints, 100Nm is selected as the biggest output of the motors. For each link, the NN has 7 hidden neurons and 1 output neuron. The learning parameters of the three NNs and initial weights are all selected as 0.5. The proportional and derivative gains of the three PD controllers are selected as $[200,1000,200]$ and $[35,80,15]$, respectively. Besides, the proportional gains and derivative gains of the three reference models are selected as $[20,100,20]$ and $[3.5,8,1.5]$ for sample 1, $[60,300,60]$ and $[10.5,24,4.5]$ for sample 2, respectively. To avoid the possible instability caused by overquick change of weights of the networks, the maximum value of s is bounded by 2. The initial conditions are $q(0)=[0,0,0]$ and $\dot{q}(0)=[0,0,0]$. Desired trajectories are selected as $q_{d1}(t)=q_{d2}(t)=q_{d3}(t)=0.5(1-\cos(\pi t))$.

Figure 2 shows the position tracking trajectories and tracking errors of the three joints by using PD controllers. Figure 3 presents the position tracking results of the three joints by using the proposed control scheme. In Figure 4, the output torque of the PD controller and the NN controller of the proposed control scheme is presented. To validate the dynamic response speed of the controller, step signal $q_{d1}(t)=q_{d2}(t)=q_{d3}(t)=1$ is applied to the controller and the output of reference models are bounded by $[-100,100]$, the results are presented in Figure 5.

The above simulation results for the three-link manipulator demonstrate that the single PD control for the three-link manipulator has low trajectory tracking precision and large position-tracking errors. This is because the PD controllers are unable to determine the appropriate PD gains in the case of nonlinear and uncertain controlled plants and can't compensate static errors. As shown in Figure 3, the proposed control scheme can realize higher trajectory tracking precision and better adaptive control capability due to the powerful capabilities of learning adaptability. Besides, as shown in Figure 5, the step response of the proposed controller is similar to PD controller. This is because the PD controller in the proposed control scheme is then used in rapid dynamic response while NN controller is used to approximate nonlinear parts with lower speed and can be deduced by Figure 4. Besides, as shown in Figure 3, 4 and 5, the trajectory errors become smaller when the gains of reference model are larger. The advantages of the proposed control scheme are that the system model does not need to be identified beforehand and the control gains can be easily tuned.

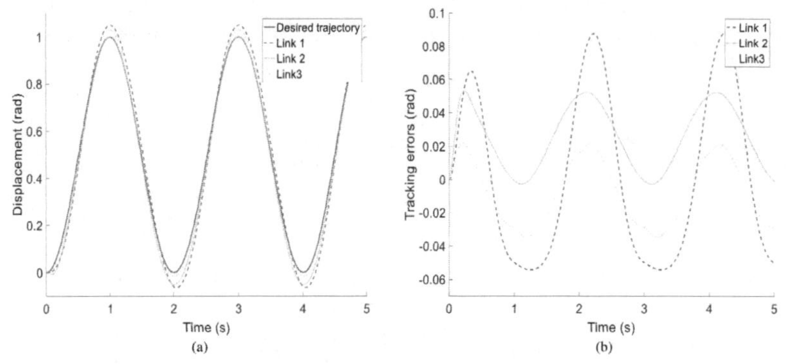

Figure 2. (a)Tracking results and (b) tracking errors of PD controllers.

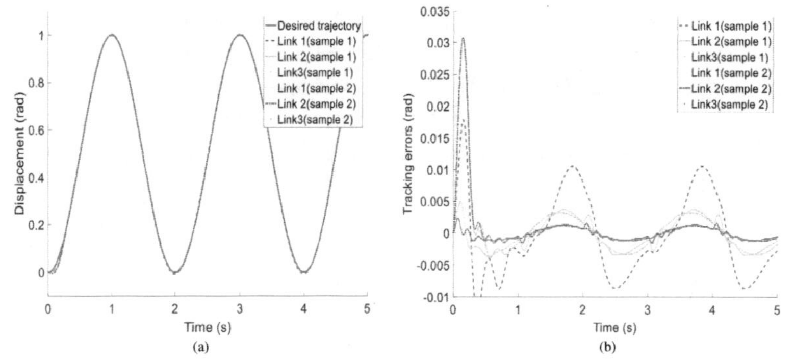

Figure 3. (a)Tracking results and (b) tracking errors of the proposed controllers.

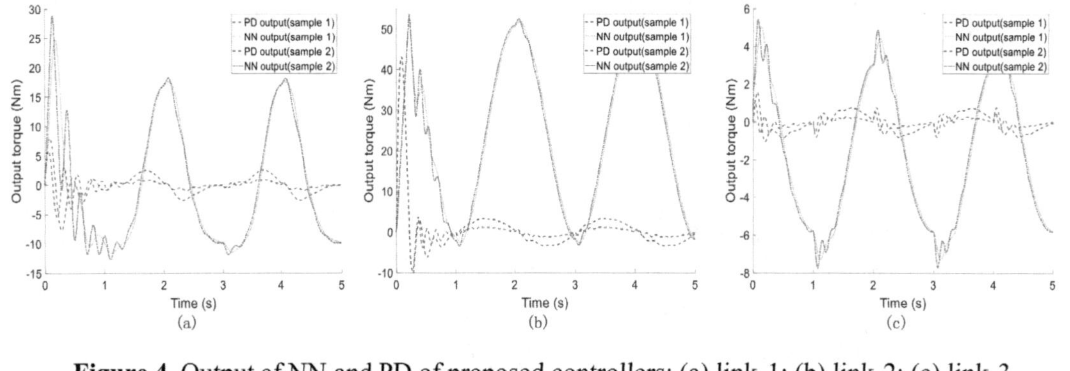

Figure 4. Output of NN and PD of proposed controllers: (a) link-1; (b) link-2; (c) link-3.

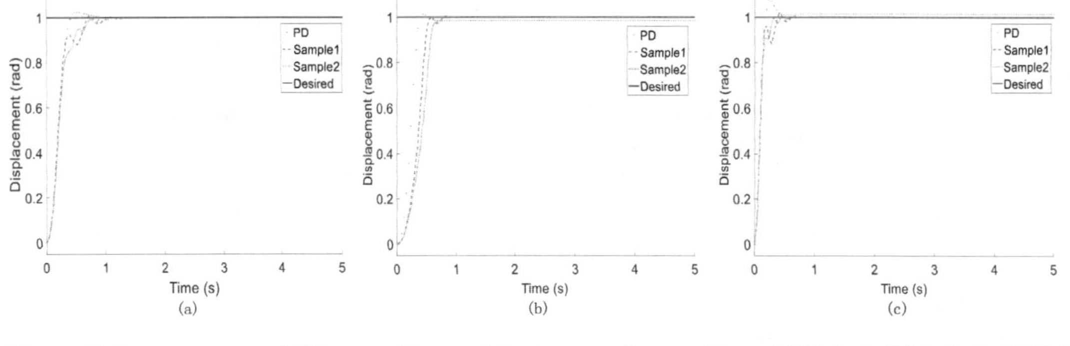

Figure 5. Step response of PD controllers and the proposed controllers: (a) link-1; (b) link-2; (c) link-3.

5. Conclusions

In this report, a new compound NN and PD control scheme has been proposed. The control scheme uses a PD controller to realize high responses speed and a neural network to approximated unknown nonlinear dynamics. A first order reference model gives the desired dynamics of the error between the desired and actual load positions. Its output is used as supplementary to PD controller and realizes approximation of unknown dynamics. Computer simulation has been carried out on a three-link manipulator with the discontinuous friction model. The results demonstrate the effectiveness of the proposed control scheme and the priorities in the parameter tuning process. Because the tracking errors of each link are the input of the proposed control scheme, distributed control can be realized. The future research work will investigate into the possibility of self-tuning methods of parameters of the proposed scheme.

6. Reference

[1] Chaoui H and Gueaieb W 2008 Type-2 fuzzy logic control of a flexible-joint manipulator *Journal of Intelligent and Robotic Systems* **51** 159-86

[2] Sun C, He W and Hong J 2017 Neural network control of a flexible robotic manipulator using the lumped spring-mass model *IEEE Transactions on Systems, Man, and Cybernetics: Systems* **47** 1863-74

[3] He W, Chen Y and Yin Z 2016 Adaptive neural network control of an uncertain robot with full-state constraints *IEEE transactions on cybernetics* **46** 620-9

[4] Ponce P, Azcue E, Silva J M, Silva J M and Fernandez R 2006 Neural Networks Based on TrigonometricSeries for the Control of a Robot. In: *Industrial Technology, 2006. ICIT 2006. IEEE International Conference on*: IEEE) pp 2302-7

[5] Hui D, Fuchun S and Zengqi S 2002 Observer-based adaptive controller design of flexible manipulators using time-delay neuro-fuzzy networks *Journal of Intelligent and Robotic Systems* **34** 453-66

[6] Chaoui H, Sicard P and Gueaieb W 2009 ANN-based adaptive control of robotic manipulators with friction and joint elasticity *IEEE Transactions on Industrial Electronics* **56** 3174-87

[7] Chen C P, Liu Y-J and Wen G-X 2014 Fuzzy neural network-based adaptive control for a class of uncertain nonlinear stochastic systems *IEEE Transactions on Cybernetics* **44** 583-93

[8] Huang S and Tan K K 2012 Intelligent friction modeling and compensation using neural network approximations *IEEE Transactions on Industrial Electronics* **59** 3342-9

[9] Rossomando F, Soria C and Carelli R 2012 Neural network-based compensation control of mobile robots with partially known structure *IET Control Theory & Applications* **6** 1851-60

[10] Le Q D and Kang H-J 2018 An Adaptive Controller with An Orthogonal Neural Network and A Third Order Sliding Mode Observer for Robot Manipulators

[11] Ye J 2014 Compound control of a compound cosine function neural network and PD for manipulators *International Journal of Control* **87** 2118-29

[12] Man Z, Wu H R, Liu S and Yu X 2006 A new adaptive backpropagation algorithm based on Lyapunov stability theory for neural networks *IEEE Transactions on Neural Networks* **17** 1580-91

7. Acknowledgments

Research supported by National Natural Science Foundation of China (Project number: 51475177) and State Key Lab of Digital Manufacturing Equipment & Technology, School of Mechanical Science and Engineering, Huazhong University of Science & Technology.

IOP Conf. Series: Materials Science and Engineering **517** (2019) 012009 doi:10.1088/1757-899X/517/1/012009

Obstacle Avoidance of Mobile Robot using Fuzzy Logic and Hybrid Obstacle Avoidance Algorithm

R Singh[1] and T K Bera[2,1]

[1 and 2] Department of Mechanical Engineering,
Thapar Institute of Engineering and Technology, Patiala, India.

Abstract. The road accidents due to traffic problems and human erroneous driving are the major challenges for researches. The self-driving car or mobile robot is the solution to avoid such mishaps. In this paper, an attempted has been made to develop obstacle avoidance algorithms for bicycle vehicle model of mobile robots. The hybrid obstacle avoidance algorithm is proposed on the merits of line, wall following and tangent bug algorithm. The trajectory generated from the hybrid obstacle avoidance algorithm is fed into the overwhelming controller of the bicycle vehicle model of mobile robot for avoiding obstacles. Then, the fuzzy logic (FL) based obstacle avoidance controller is proposed. Twenty-three set of rules are proposed for fuzzy logic approach. Both the obstacle avoidance algorithms are implemented on the bicycle vehicle model of the mobile robot. The dynamic model of the mobile robot is developed using bond graph theory and is converted into *Simulink* block using *S-function* directly from the library of *SYMBOLS Shakti* software. The vehicle model is equipped with three ultrasonic sensors to measure the distance from the obstacles. Three input membership functions and one output membership function are considered in fuzzy logic controller. During avoiding two static obstacles and reaching the target, the comparison of obstacle free paths traced by both obstacle avoidance algorithms is done in this paper.

Keywords: - Hybrid obstacle avoidance, fuzzy logic controller, mobile robot, bond graph.

1. Introduction

The majority of the accidents occur on the road because of human harsh driving. The solution to solve this dilemma is the development of the self-driving vehicle. A self-driving vehicle is able to sense its environment and navigate to the target without human assistance. It has capability to reduce mishaps due to driver's fault [1]. A controller is developed for analyzing the dynamics of vehicle [2, 3]. The Automated guided vehicle (AGV) is also helpful in various tasks. The combination of fuzzy and Vector field histogram (VFH) is also suitable to bypass the obstacles [4]. Jun *et al.* used two Light detection and ranging (LIDAR) sensors for obstacle avoidance of mobile robot [5]. The use of geometry modelling for object avoidance approach is mentioned in [6]. The fuzzy controller may be used for local and global navigation and steering control of the robot [7]. Xiong *et al.* used two FL controllers to control the intelligent vehicle [8]. The modified artificial potential field method to generate trajectory for wheeled mobile robot is presented in [9]. To find shortest path with obstacle avoidance for wheeled mobile robot based on fuzzy, ant colony optimization technique are proposed in

[1] **Corresponding Author**: - Dr. Tarun Kumar Bera
Department of Mechanical Engineering,
Thapar Institute of Engineering & Technology, Patiala, 147004, India,
Email: tkbera@thapar.edu, tarunkumarbera@gmail.com

[10]. The wheel mobile robot using arduino board with ultrasonic sensor and compass is developed for avoiding static obstacles [11]. Predicting the movement trends of the obstacle using movement laws of the dynamic obstacle is proposed for mobile robot to plan the obstacle free path [12].

The bond graph (BG) is a useful technique to model complicated systems. BG approach is extensively applicable in vehicle dynamics [13]. BG is used to study the bicycle vehicle model for obstacle avoidance using overwhelming controller [14]. In this work, the hybrid Obstacle avoidance algorithm (OAA) is proposed using line, wall following and tangent bug algorithm. Then, FL based OA controller is proposed. A set of 23 rules are created on the basis of three inputs and one output membership functions. After that, the BG model of Bicycle vehicle model (BVM) is build using *SYMBOLS* Shakti software. The development of the hybrid obstacle avoidance algorithm by using the merits of line following, wall following and tangent bug algorithm is a new work. This algorithm is used to follow the shortest possible path based on the geometry of obstacle by following its circumferential boundary. Also, the fuzzy logic based OAA is useful in identifying not only the position of the robot but also, it is capable to determine the orientation of the robot w.r.t. the target. This is the novelty of the work. The work is organized in the following sections: First, the hybrid obstacle avoidance and fuzzy based algorithms are proposed. Then, BG model of BVM is developed and converted into *Simulink* block. After this, both the obstacle avoidance algorithms are implemented on the BVM and result regarding obstacle free path is compared. Finally, the conclusions are drawn.

2. Obstacle avoidance algorithm (OAA)

2.1 Hybrid obstacle avoidance algorithm
A hybrid OAA is proposed to avoid the static objects by a biped robot. The hybrid obstacle avoidance algorithm comprises of utilizing the advantages of line following algorithm, wall following algorithm and tangent bug algorithm for avoiding static obstacles while reaching the target by the biped robot. In this algorithm, assumptions of the pre-defined location and geometry of obstacle along with the nature of target and obstacle being static are considered. The flow diagram of hybrid OAA is depicted in Fig. 1(a). Initially, the robot follows the predefine path using line following algorithm. If any object is sensed by ultrasonic sensor and if the distance between the robot and the object is less than the specified distance, tangent bug algorithm comes into the picture. Using image-processing technique, the current pose and orientation of the robot is calculated from the data sensed by the overhead camera. These data are fed to the control unit which calculates the minimum distance of the robot to reach the target by avoiding the obstacle. Then, the robot follows the wall following algorithm maintaining safe distance from the obstacle. After the obstacle is avoided, the robot again follows the predefined path to reach the target. The detail flow chart of OAA is show in Fig. 1(b).

2.2 Fuzzy logic obstacle avoidance algorithm
The fuzzy logic approach for obstacle avoidance is proposed. The inputs to the controller are the distances from the target and the obstacle, heading angle. The controller output is the steering angle of the vehicle. The block diagram of fuzzy based obstacle avoidance controller is depicted in Fig. 2. A set of 23 rules are proposed to avoid the static and dynamic obstacles by the four-wheel mobile robot. Three ultrasonic sensors are mounted on the head of the robot to capture the data i.e. distance from the obstacles. Figure 3 shows the fuzzy controller input and output membership functions.

3. Bond graph model of bicycle vehicle model
Bicycle vehicle model is half model of the full vehicle model and helps in studying the vehicle behavior. The heave, roll and pitch motions and suspension dynamics are not considered during modelling. Line diagram of the BVM is illustrated in Fig. 4(a) and (b) show its word BG. The steering angle δ is decided the turning of the front wheel. The power is given to rear wheel. Normal and tangential velocities for front and rear wheels and Newton-Euler equations are required to complete bond graph model of bicycle vehicle and this model is illustrated in Fig. 4(c). The BG model is converted into the Simulink block which is further used in the *MATLAB* software for implementation of the OAA.

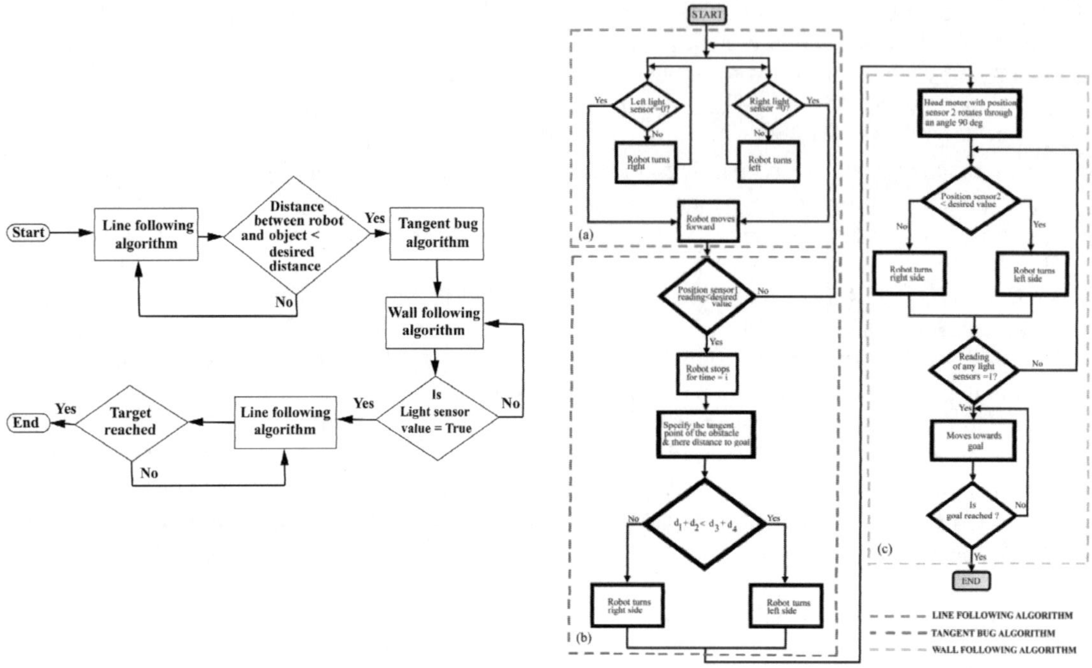

(a) (b)

Figure 1 (a) Flow diagram and (b) Flow chart of obstacle avoidance algorithm [10]

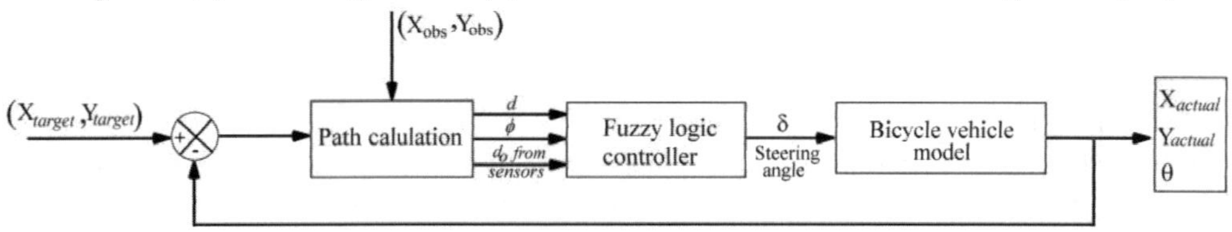

Figure 2 Block diagram for navigation and obstacle avoidance of bicycle vehicle model

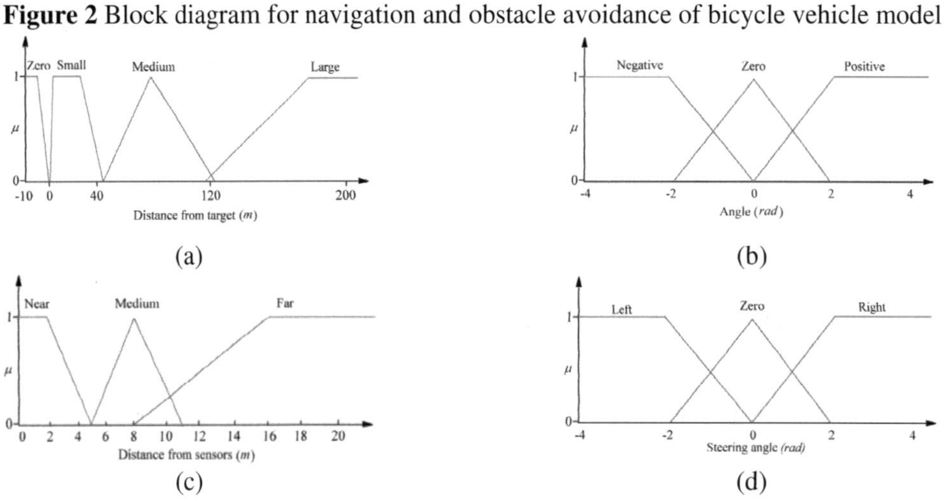

(c) (d)

Figure 3. The fuzzy membership functions (a-c) Input and (d) Output

4. Simulation results

The performance of the hybrid OOA and fuzzy logic OAA using bicycle vehicle model is proposed. Table 1 shows the case considered under simulation work. In this scenario the two circular static objects are considered in the path of the bicycle vehicle model from start to target point. The location of the object is depending upon the position of the object centroid w.r.t. to center line. Table 2 shows the simulation parameters.

IOP Conf. Series: Materials Science and Engineering **517** (2019) 012009 doi:10.1088/1757-899X/517/1/012009

Table 1. Case for simulation

Cases	No. of object	Position of object centroid		Geometry
Case-I	Two	Left of center line	Right of center line	Circular shape

Table 2. Simulation parameters

Parameters	Description	Value
r_W	Wheel radius	0.254 m
m_V	Mass	924 kg
a	CG to front wheel distance	1.31 m
b	CG to rear wheel distance	0.62 m
k_P	Spring stiffness	10^8 N/m
r_P	Damping stiffness	10^8 Ns/m

The comparison of hybrid obstacle avoidance and fuzzy logic obstacle avoidance controller for avoiding two static obstacles is shown in Fig. 5. As depicted from the Fig. 5 the fuzzy logic controller generates the object free path, which is close to the obstacles as compared to the hybrid obstacle avoidance algorithm. Hybrid OAA maintains safe distance from the obstacles. So, Fuzzy logic obstacle avoidance algorithm may be used when safe distance is not important but time to complete the path is more significant.

5. Conclusions

The obstacle avoidance algorithms were developed and implemented on bicycle vehicle model. By using line, wall following and tangent bug the hybrid obstacle avoidance algorithm was developed. Also the fuzzy logic obstacle avoidance algorithm was proposed. Three inputs and one output membership functions are developed for the fuzzy controller. The bond graph approach is used to develop the bicycle vehicle model. This model is then converted into the *Simulink* block. The practicality of both algorithms has been compared for two static obstacles in bicycle vehicle model. The fuzzy logic obstacle avoidance algorithm generates the object free path, which is close to the obstacles as compared to the hybrid obstacle avoidance algorithm.

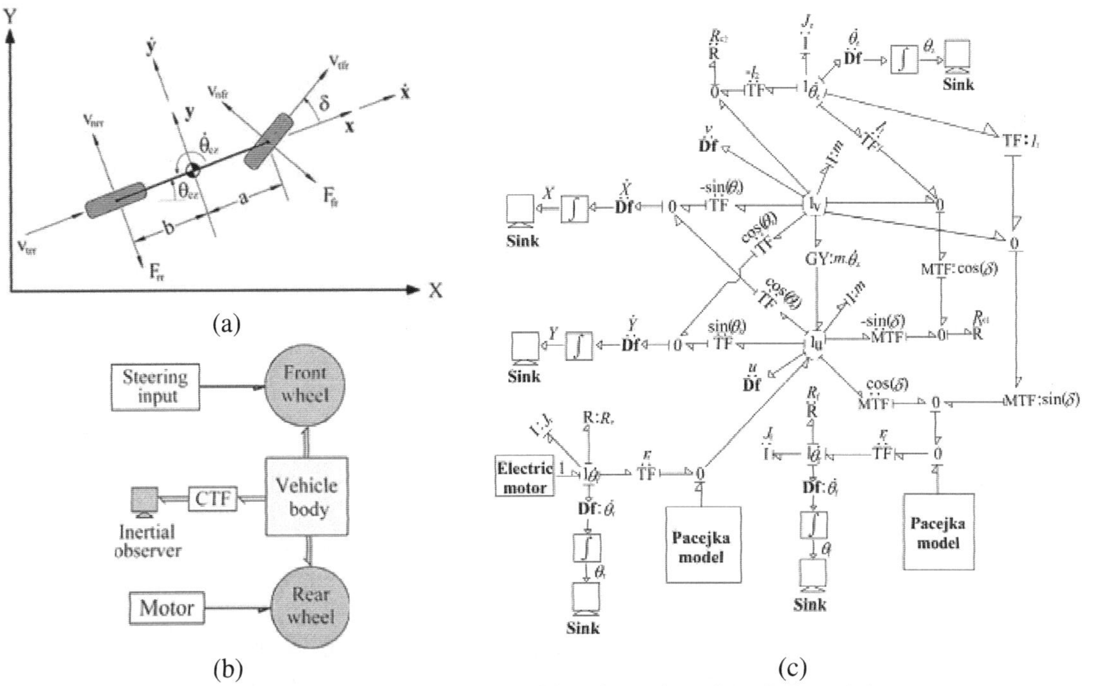

Figure 4. (a) Graphical representation, (b) Its word bond graph and (c) BG model

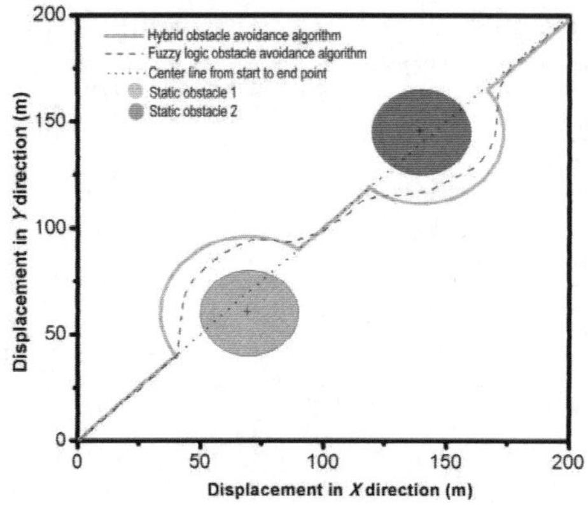

Figure 5. Comparison of obstacle free path using hybrid and fuzzy logic obstacle avoidance algorithm

6. References

[1] Iida F, Gomez G, Pfeifer R 2005 Exploiting body dynamics for controlling a running quadruped robot *Proc. Int. Conf. Adv. Robo. ICAR* (USA) p 229

[2] Thrun S, Montemerlo M and Dahlkamp H 2006 The robot that won the DARPA grand Challenge. *J. of Field Robo.* **23** 661

[3] Linhui L, Mingheng Z, Lie G and Yibing Z 2009 Stereo Vision based Obstacle Avoidance Path-planning for Cross-country Intelligent Vehicle. *Proc. In: IEEE International Conference on Fuzzy system and knowledge Discovery* (China) p 463

[4] Ghoshray S and Yen KK 1996 A Comprehensive Robot Collision Avoidance Scheme by Two Dimensional Geometric Modeling. *Proc. In: IEEE International Conference on Robotics and Automation* (Minnesota) p 1087

[5] Jun J, Zhan WC and Juang J 2010 Application of Image Process and Distance Computation to WMR Obstacle Avoidance and Parking Control. *Proc. In: IEEE Conference on Industrial Electronics and Applications* (Taiwan)p 1264

[6] Park CW and Lee CY 2010 Sonar sensor data processing based on optical flow in robot navigation. Proceedings of the Institution of Mechanical Engineers, *Part 1: Journal of Systems and Control Engineering* **225** 1

[7] Moon AP and Jajuiwar KK 2016 Design of adaptive fuzzy tracking controller for Autonomous navigation system, *Int. J. of Recent Trend in Engg. and Research* **2** 268

[8] Xiong B and Qu SR 2010 Intelligent vehicle's path tracking based on fuzzy control, *J of Transp Syst Eng and Inf.* **10** 70

[9] Sudhakara P, Ganapathy V, Priyadharshini B and Sundaran K 2018 Obstacle Avoidance and Navigation Planning of a Wheeled Mobile Robot Using Amended Artificial Potential Field Method. *Procedia Computer Science.* **133** 998

[10] Yen CT and Cheng MF 2018 A Study of Fuzzy Control with Ant Colony Algorithm used in Mobile Robot for Shortest Path Planning and Obstacle Avoidance. *Microsystem Tech.* Springer **24** 125

[11] Mohomed YA and Ali HA 2018 Wheeled Mobile Robot Obstacle Avoidance using Compass and Ultrasonic. *Universal J. of ctr. and autom.* **6** 13

[12] Liu Y, Chen D and Zhang S 2018 Obstacle Avoidance Method Based on the Movement Trend of Dynamic Obstacles. *In.: Proceeding of 3rd International conference on control and robotics engineering.* (Japan) p 45

[13] Bouamama B, Medjahar K, Bayart M, Samantary A K and Conrard B 2015 Fault detection and isolation of smart actuators using bond graph and external models, *Ctr. Eng Pract,* **13** 159

[14] Singh R and Bera T K 2018 Obstacle avoidance of bicycle vehicle model using overwhelming controller, Arabian *J Sc Eng.* **43** 4821

IOP Conf. Series: Materials Science and Engineering **517** (2019) 012010 doi:10.1088/1757-899X/517/1/012010

Evaluation of Noise Performance of Multi-Lane Highways in the State of Qatar

Okan Sirin[1], Md Ohiduzzaman[2], Emad Kassem[3]

[1] Department of Civil and Architectural Eng., Qatar University, P.O. Box 2713, Doha, Qatar

[2] High Definition Design PTY LTD, 16 Russel Street, Cardiff, Newcastle, NSW-2285 Australia

[3] Department of Civil and Environmental Eng., University of Idaho, Moscow, Idaho 83844 USA

Abstract. In recent years, a growing number of people are complaining to transportation authorities for the traffic noise as a result of increased traffic volume on the highways. Therefore, transportation authorities are actively pursuing ways to reduce the traffic noise. In recent years, quieter pavement has become one of the cost-effective ways of sustainable highway noise reduction. However, for design and construction of quieter pavements, proper and accurate noise measurement is of paramount importance for evaluating the noise level of existing pavement as well as evaluating noise level of newly constructed pavements. In this study, the variation of noise level among different lanes of existing multi-lane highways in the State of Qatar was examined. Tire-pavement noise was measured by using onboard sound intensity (OBSI) technique. OBSI test results showed that highest noise level was observed at outer lane of multi-lane highways. This is probably due to the pavement distress caused by the heavy vehicle travelling at the outer lane of multi-lane highways.

1. Introduction

Noise pollution due to moving traffic has been one of the important environmental concern for the people living close to highway vicinity. A number of psychological and health related problems of highway inhabitants has been linked to traffic noise [1]. Therefore, transport agencies are actively finding ways to minimize the noise from the road traffic. Although traffic noise is comprised of number of sources, but tire-pavement interaction noise is the major source of traffic noise at medium to higher vehicle speed [2, 3]. Therefore, reducing tire-pavement noise at source could be an important option in order to balance the benefits of well-connected highway network against the adverse impact of the residents exposed to highway traffic.

A number of ways to reduce the tire-pavement noise was adopted by various transportation authorities. Of all the noise reducing ways, three techniques are generally used which included producing quieter tire, constructing quieter pavement and providing a barrier wall between source and receiver [2]. Tire manufacturers are actively pursuing to produce tires which minimize the impact between tire and pavement. Although, noise barrier wall is most commonly used by transportation authorities, but they are not always effective due to practical problems and are very expensive (~$2.1 million per mile) [4]. On the other hand, worldwide transportation authorities have been providing great effort to identifying ways of noise reduction by altering/optimizing the road surfaces properties. However, there is always a concern that optimizing pavement surface could compromise other attributes of pavement such as safety,

IOP Conf. Series: Materials Science and Engineering **517** (2019) 012010 doi:10.1088/1757-899X/517/1/012010

durability and cost of pavement. Nevertheless, researchers in Europe and USA showed that constructing quieter pavement is possible within a reasonable cost without compromising safety and long-term durability of pavement [2].

Accurate measurement of tire-pavement interaction noise is required because the data which are collected from the measurement system will be used to identify quieter pavement and to understand the noise generation mechanisms. At present, noise measurement in the field is generally performed in two ways: wayside noise measurement and noise measurement at source. However, if the goal is to identify quieter pavement, source noise measurement is more accurate compared to wayside noise measurement [2, 3]. The source noise measurement method provides a direct measure of noise in close proximity to the tire-pavement interface and allows various pavement types and textures to be directly compared [2, 3]. The main objective of this study was to examine the noise performance of various lanes of multi-lane highways in order to develop guidelines for the sustainable quieter pavement in the State of Qatar.

2. Experimental Program

2.1. Locations of Test Sections

Researchers selected pavement sections on two major highways (i.e., Dukhan and Al Shamal) in the State of Qatar to identify the variation of noise level in different lanes of multi-lane highway. Both of the highways have dense graded asphalt concrete (DGAC) surface which represents most of the pavement sections in the State of Qatar and constructed by using gabbro aggregates. Noise testing was performed at number of sections as both highways are very long. To investigate noise performance of various lanes, noise testing was performed at three lanes of both highways. At first, noise testing was conducted at outer lane of highway and followed by inner lane 1 and 2. Figure 1 depicts the locations of the lane on the map for Al Shamal highway.

Fig. 1. Lane locations for Al Shamal highway

2.2. Noise Measurement Method

In this study, tire-pavement noise was conducted on selected test sections by using onboard sound intensity (OBSI) method following AASHTO T 360-16 [5] standard. In this method, sound intensity is measured by placing sound intensity probes consisting of pair of microphones close to the tire-pavement interface. Two sound intensity probes are placed at specified positions (i.e., 101.6 mm from the center of the tire and 76.2 mm above the surface of the pavement) in accordance with AASHTO T 360-16 standard [5]. Fig. 2 shows the instrumented OBSI set up on the field test section. All the OBSI testing were conducted at AASHTO T360-16 [5] specified speed of 96.6 kM/hr (60 mph) with a tire inflation pressure 30 psi in tire cold condition. Meteorological data was also recorded during noise measurement as it impacts the generation and amplification mechanism of tire-pavement interaction noise [2, 6].

Fig. 2. Instrumented OBSI equipment on field testing section

3. Discussion of Test Results

OBSI noise testing was conducted on each section in accordance with AASHTO T 360-16 [5] standard. At least three valid runs were performed in each test section as there is slight variation of noise level among successive runs even though test was conducted by the same equipment and same operator. During post processing of data, the average noise level was calculated which represents the noise level of that particular section. Furthermore, noise level of different sections in any specific lane was varied due to slight variation of pavement during construction period and distress caused by the moving traffic due to operational period [2, 6, 7]. However, the variation of measured overall intensity level among sections in any lane is small as observed from the test data. Therefore, noise level of all the sections of any specific lane were averaged to create a representative noise profile for that highway lane. The test results for both highways at three respective lanes are shown in Fig. 3. It can be seen from figure 3 that noise intensity level is highest for outer lane and lowest for inner lane 2 for both highways (i.e., Dukhan and Al Shamal). Almost 3 dBA noise level difference was observed between outer lane and inner lane 2 for both highways. Most of the heavy vehicles are generally traveled along the outer lane of multi-lane highways which may resulted more distress and wear in pavement hence showed higher noise level. However, texture measurement at different lane is required to verify this hypothesis.

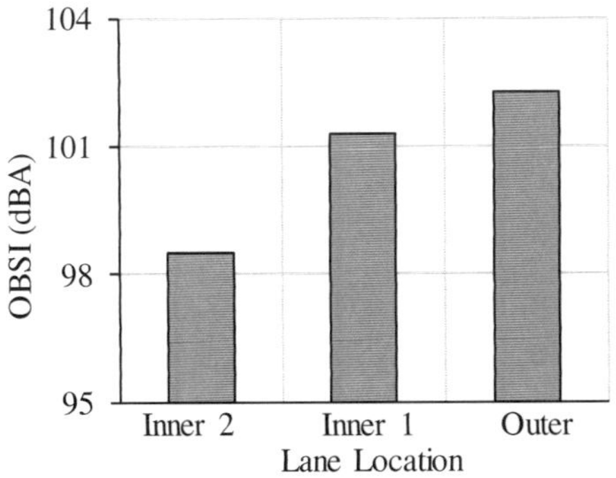

a)

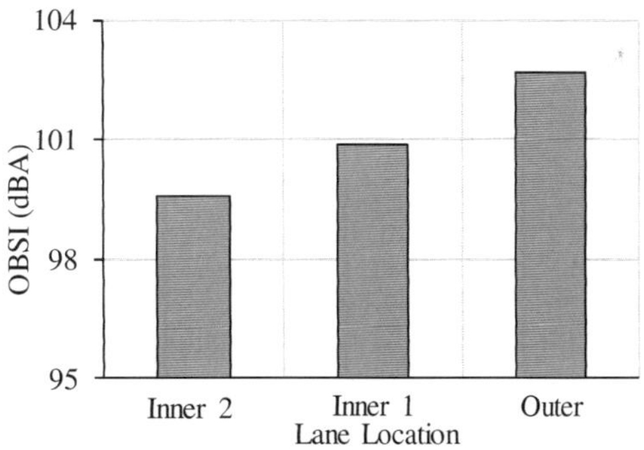

b)

Fig. 3. OBSI data for pavement sections; a) Dukhan Highway; b) Al Shamal Road

To identify the mechanism that affects the acoustical behavior of different lane of highway, the frequency response of test sections was compared and shown in Fig. 4 and Fig. 5 for Dukhan highway and Al Shamal road, respectively. The peak sound intensity amplitude occurred at the center frequency of 1000 Hz frequency band irrespective of lane location for both highways. This demonstrated that the prominent peak around 1000Hz for small car noise spectra for DGAC surface and is also consistent with the literature [8, 9]. The noise intensity of the outer lane is lying well above compared to the other two inner lanes at all frequency range for both highways. At a lower frequency below 1000 Hz, the increase of noise intensity for outer lane of pavement was more prominent. This indicates some changes in surface texture of pavement due to wear and tear caused by heavy vehicle which generally travel on outer lane of multi-lane highway. Tire vibrations due to high macrotexture dominants tire-pavement the at the low frequencies whereas tire/pavement noise were found to be governed by air-pumping mechanisms at higher frequency that can be reduced by the presence of air voids on the pavement surface [2].

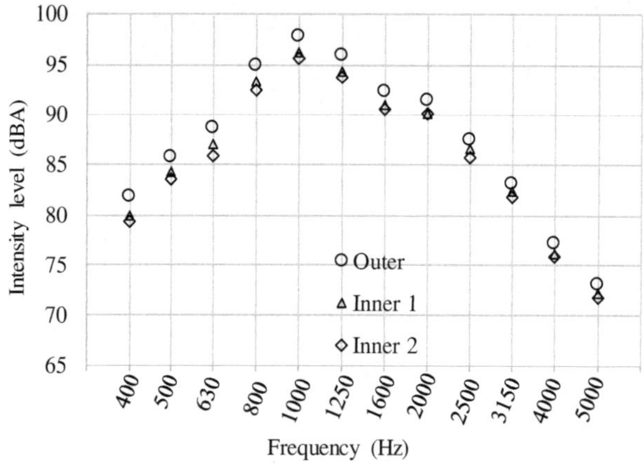

Fig. 4. Frequency response of various lanes at Dukhan Highway

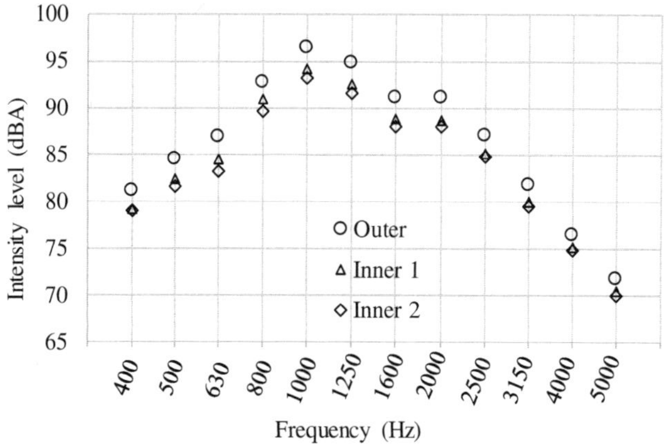

Fig. 5. Frequency response of various lanes at Al Shamal Road

4. Conclusions

Tire-pavement noise generated due to interaction between rolling tire and pavement surface can be reduced by proper design of pavement surfaces. In this study, noise performance of various lanes of two major highways in the State of Qatar was investigated by using OBSI technique. OBSI test results showed that noise level of outer lane was generally the loudest for both highways, resulting in overall OBSI levels of about 3 dBA differences between outer and inner lane of highway. This study provides insights the acoustical performance of asphalt mixture subjected to real traffic and harsh climate conditions which will be important when preparing quieter pavement guidelines in the State of Qatar.

5. Acknowledgments

This paper was made possible by the NPRP grant (NPRP 7-110-2-056) from the Qatar National Research Fund (a member of Qatar Foundation). The statements made herein are solely the responsibility of the authors.

6. References

[1] WHO. Night noise guidelines for Europe. Regional Office for Europe, Copenhagen, Denmark:2009.http://www.euro.who.int/__data/assets/pdf_file/0017/43316/E92845.pdf, accessed April (2018).

[2] Ohiduzzaman, M., Sirin, O., Kassem, E. and Rochat, J. "State-of-the-Art Review on Sustainable Design and Construction of Quieter Pavements—Part 1: Traffic Noise Measurement and Abatement Techniques" Sustainability (MDPI), Vol 8 Issue 8 (2016), 742.
[3] Sandberg U, Ejsmont J. Tire/road noise reference book. Informex, Kisa, Sweden: (2002).
[4] Hanson, D.I., James, R.S., NeSmith, C. "Tire/Pavement Noise Study." National Center for Asphalt Technology, Report 04-02 (2004).
[5] AASHTO TP 360-12-Standard method of test for measurement of tire-pavement noise using the on-board sound intensity (OBSI) method. American Association of State Highway and Transportation Officials (AASHTO), Washington, D.C (2016).
[6] Rasmussen, R., Bernhard, J., Sandberg, U. & Mun, P., The little book of quieter pavements. The Transtec Group, Inc., Austin, Texas (2007).
[7] Sirin, O. State-of-the-Art Review on Sustainable Design and Construction of Quieter Pavements—part 2: factors affecting tire-pavement noise and prediction models. Sustainability (MDPI), Vol 8 Issue 7 (2016), 692.
[8] Ohiduzzaman, M., Sirin, O., Kassem, E. Assessment of tire-pavement noise by using on-board sound intensity (OBSI) method in the State of Qatar. 10th International conference on the bearing capacity of roads, railways and airfields (BCRRA 2017), JUNE 28~30, Athens, Greece, (2017).
[9] Sandberg, U. The multi-coincidence peak around 1000 Hz in tire-road noise spectra. Euro noise, Paper ID: 498, Naples, (2003).

IOP Conf. Series: Materials Science and Engineering **517** (2019) 012011 doi:10.1088/1757-899X/517/1/012011

A Supervised Learning Approach to Appliance Classification Based on Power Consumption Traces Analysis

J R Cuñado[1] and N B Linsangan[2]
[1] Department of Engineering Education, University of Mindanao – Tagum, Visayan Village, Tagum City, Davao del Norte 8100, Philippines
[2] School of EECE, Mapua University, 658 Muralla St., Intramuros, Manila 1002, Philippines

Abstract. Electric appliances are everyday major power consumers. Management and control of these appliances can only be possible with appliance classification infrastructures. An appliance classification smart meter, with a provision for remote control, is developed based on time-dependent power features drawn by an appliance, from power-up to its steady state. The kNN classifier is highly accurate at 95.55% in classifying the appliance class.

1. Introduction

The role that electric appliances play in enhancing the living styles of consumers can no longer be discounted at this time and age. With these devices central to the increasing everyday power consumption in both homes and offices, power consumption management is of growing interest to reduce electricity bills. Also, instances of neglect to turn off appliance power which instigated fire had been reported several times. The need, therefore, to provide a means to connect and control electric appliances remotely is seen to be of timely importance.

Appliance Load Monitoring (ALM) has been the subject of recent researches. Plug load identification is tricky, with each appliance distinct, depending on the model, type and brand [1]. Two existing approaches are used in identifying appliances based on power consumption traces: Non-Intrusive Load Monitoring (NILM) and Intrusive Load Monitoring (ILM). Hart introduced the NILM approach where a meter-extension collar containing Non-Intrusive Load Monitor is used to check for load signatures to infer appliances' activities. He based his unique appliance signature on the real and reactive power consumed by the each appliance in steady state [2]. With NILM, aggregate electricity consumption is measured at a central point using a single sensor; disaggregation is then done to determine the contribution of individual appliance. The use of decentralized meters to measure power consumption at appliance level is a relatively new research interest, with researchers looking for better ways to manage and control household devices. Ito was one of the first researchers who undertook a study in this field by matching the feature parameters of current of the appliance to the feature parameters stored in a database [3]. Reinhardt improved on Ito's work by extracting current-based and harmonics features from medium to higher frequency data [4]. High-frequency measurements of voltage and current can be used to build a coarse-grained model to detect attached appliance type and a fine-grained model to differentiate between products of an appliance type [5]. However, the use of medium- to high-frequency sampling is resource demanding, the reason why other researchers sampled data at low frequency using low-end sensors with promising results in appliance identification. Ridi further added a User Interaction Layer in the context of appliance monitoring [6]. Machine learning approaches have been used for appliance identification, and most are based on supervised learning techniques [7]. Hidden Markov Models outperformed Dynamic Time Warping on a 6-category task [8]. One-class Support Vector Machines with Gaussian Kernels achieved a 99.9% accuracy for MS-ILM [9]. kNN systems' performance on a 2-class task is 87.5% [10]. The accuracy of kNN for category recognition for signatures sampled at 4.4 kHz was at 85.5% [11]. The choice of machine learning algorithms in appliance classification however varies with appliance categories and with different types of measurements. Recent researches proposed load monitoring systems based on different frameworks, e.g. using Interactive Voice Response [12], IoT [13], and a home energy saving network [14], and Murali proposed load monitoring via social media [15]. The scope, however, as to accuracy and specific appliance types and feature sets considered, especially those that used machine learning approaches, is not clear.

Most commercial energy meters do not come with appliance identification features which can be useful for remote load monitoring and control. With ILM a recent research area of interest, publicly available datasets for ILM are scarce, with signatures recorded at low frequency and with very limited appliance types, influencing significantly the reported performance of scientific works. To date, no ILM study has been conducted yet in the Philippines. This study explores the concept of developing a system for appliance classification using a supervised learning technique based on power consumption traces following the ILM approach.

The main objective of the study is to develop a system for appliance classification based on power signatures. Specifically, the study aims to: (1) develop an individual appliance metering device to be used in appliance classification; (2) identify appliances based on power signatures using a supervised learning technique; and (3) create a load library of appliance power signatures as corpus data for training and recognition.

With the availability of an appliance classification system, with provisions to connect and control electric appliances remotely, acquisition of data can be made convenient for monitoring an appliance's consumption and usage patterns which can be used in managing appliance usage. Also, hazards caused by neglect to turn off appliance power which may lead to fire can be avoided.

Five appliance types are considered in this study, namely, computer set, electric fan, laptop (via charger), printer, and mobile phone (via charger). The choice of the appliance to be included in the study is based on the available appliances common to ten offices in UM Tagum College. This study covers the use of time-dependent power features drawn by an appliance, from power-up to its steady state, as bases for its classification. Also, this study assumes that the appliance to be subjected for classification is in good working condition.

2. Methodology

2.1. System architecture

The architecture of the appliance classification system is shown in figure 1. The smart meters are attached between the home outlets and the appliance. These meters perform data acquisition, feature extraction and appliance classification, in that order. A user interface (computer monitor) can be used to access a meter for maintenance purposes. The load library of appliance power signatures to be used for classification is local to the meter. Classification results are relayed to the server through the gateway, which contains drivers and protocols to enable the meter to communicate over the public Wide Area Network – the Internet. The smart meters have cloud-based back-end to provide information for views using a mobile device. The mobile device can control the turning on or off of the connected appliance, allowing not only remote user access but also control.

2.2. Smart meter hardware

As shown in figure 2, the appliance is connected in parallel with the main voltage source. The potential transformer (PT) and the current transformer (CT) step down the high value voltage and current levels, respectively, to safe values compatible with digital circuitries. The Raspberry Pi is used as the central processing unit of the smart meter because of its low-cost and reliability, appropriate for wireless sensor network monitoring system [16]. But since the Raspberry Pi does not come with a built-in analog-to-digital converter, a high precision Analog-to-Digital Converter (ADC) is used to convert the sensed analog voltage and current values to their digital equivalents. The latching relay acts as an actuator to turn on or off a plugged appliance when a control signal from the Raspberry Pi is received. The Raspberry Pi has a built-in wifi module to enable the meter to communicate to a mobile device via the Internet.

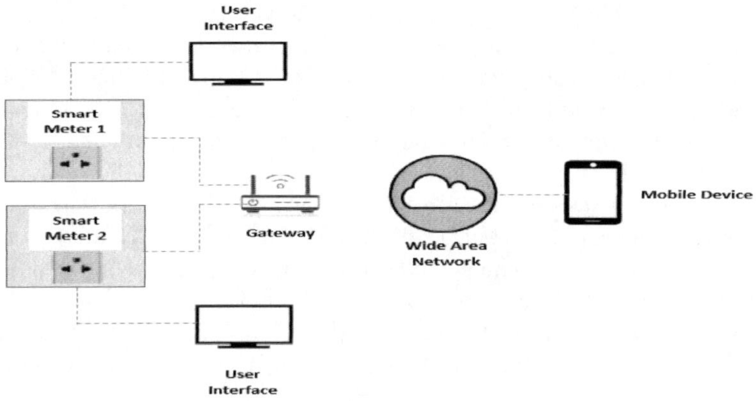

Figure 1. Appliance classification system architecture.

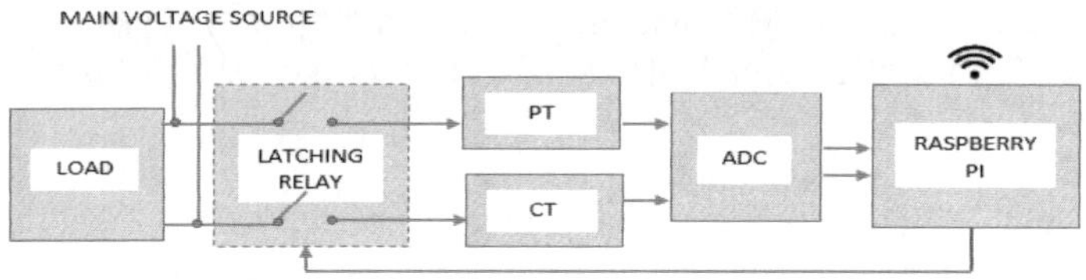

Figure 2. Smart meter hardware block diagram.

2.3. *Appliance classification framework*

Appliance classification seeks to determine the class of a new observation, based on a training data set containing observations whose classifications are known. The framework, whose major components are discussed below, is used in this study as shown in figure 3.

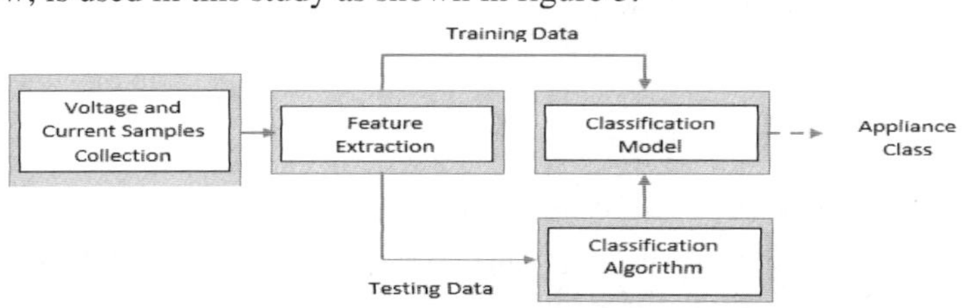

Figure 3. Appliance classification framework.

2.3.1. Voltage and current samples collection.

Data acquisition is crucial in classification problems in that the results of the succeeding processes are greatly dependent on the soundness of the collected data. For data involving continuous signals, the timing and length of collection are critical. The smart meter developed for the purpose of this study is used to acquire voltage and current readings, with the appliance initially turned off. Seventy five plug loads from five appliance classes, fifteen of each class, including computer set, electric fan, laptop (via charger), printer, and mobile phone (via charger) were recorded for the training data set. Two thousand samples per appliance class were collected at a rate of 100 samples per second. The data were collected from the time the appliance was turned on until its steady state. Data for appliances via chargers were taken while the appliance was charging.

IOP Conf. Series: Materials Science and Engineering **517** (2019) 012011 doi:10.1088/1757-899X/517/1/012011

2.3.2. Feature extraction. Feature extraction is the process of deriving useful and distinct quantities from an initial set of measured data. In the context of electrical metering devices, the voltage and current readings obtained from the analog-to-digital converter were in the form of counts. These data still need to be pre-processed, based on the ADC's resolution and the system voltage and current, to determine the analog voltage and current values measured, respectively. These values are then used to derive relevant electrical quantities to form part of an appliance's power signature. The feature set used in this study includes real power (P), peak current (I_{PEAK}), RMS current (I_{RMS}) and the power factor (PF). Frequency and RMS voltage are not included since these quantities are considered constant for all devices and so are considered redundant features.

2.3.3. Classification model. The classification model is the library of appliance power signatures, appended with a corresponding appliance classification label for each signature set that will be used by the classifier to predict the appliance class.

2.3.4. Classification algorithm. The k-Nearest Neighbors classifier works by comparing the most similar piece of data to the training set and returns the class label [17]. The k value is the number of most similar pieces of data from the training set to be resolved by a majority vote. The classifier takes in four inputs: the input vector to classify, the matrix of training data set, the vector of labels and the value of k nearest neighbors to use in voting. To deal with values that lie in different ranges, normalization is done.

3. Results and discussions

The smart meter developed for this study was used for data gathering. After collection and the data pre-processed, the current values were plotted to visualize the actual current drawn by an appliance at start-up until its steady state for analysis.

The time it takes for the current to get to steady state differs from one appliance class to another. The computer set took approximately eight seconds after turning on to get to its steady state, and its current peaked, on average, at 0.464569155 A, as shown in figure 4.

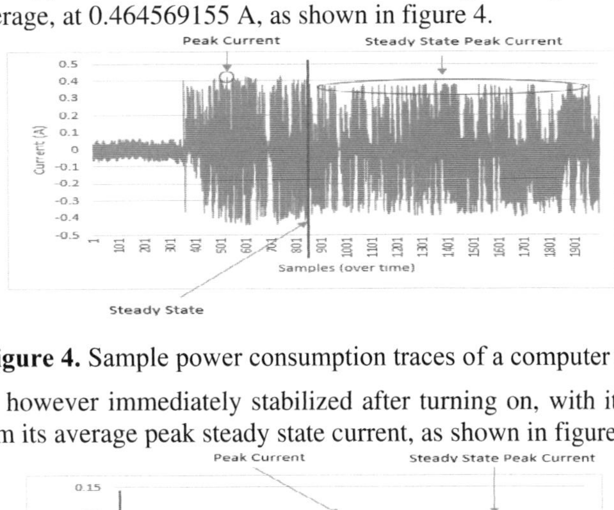

Figure 4. Sample power consumption traces of a computer set.

The mobile phone however immediately stabilized after turning on, with its peak current vary by only about 0.01 A from its average peak steady state current, as shown in figure 5.

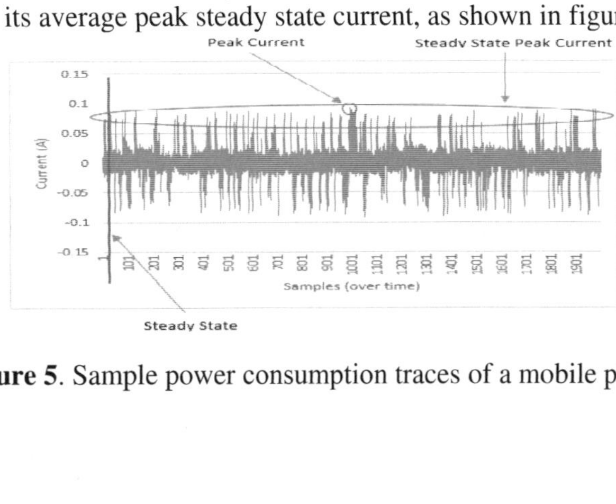

Figure 5. Sample power consumption traces of a mobile phone.

The electric fan peaked, on average, at 0.343963255 A and took approximately three seconds to get to steady state, as shown in figure 6.

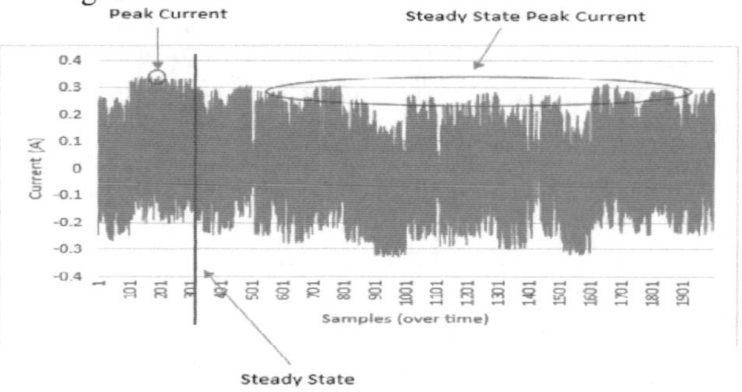

Figure 6. Sample power consumption traces of an electric fan.

In figure 7, it can be observed that the laptop current steadied after about two seconds from turning on and peaked at 0.46006135 A.

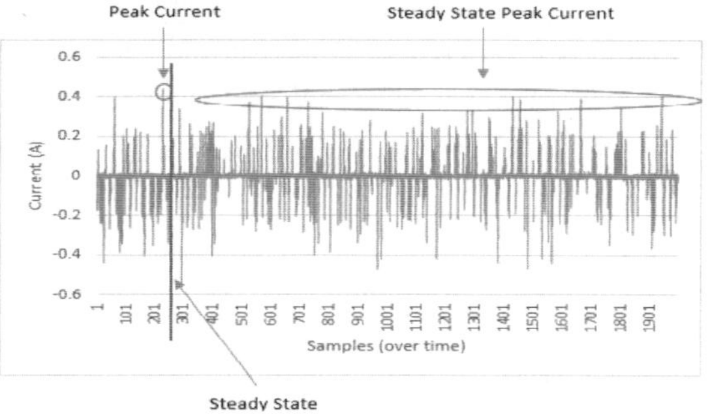

Figure 7. Sample power consumption traces of an electric fan.

Very noteworthy is the large start-up current drawn by the printer shown in figure 8 which peaked at 3.704620815 A and significantly dropped approximately to almost 0 A eight seconds after turning on.

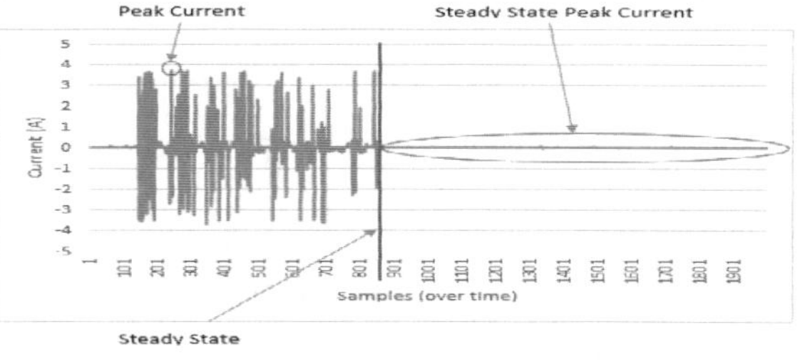

Figure 8. Sample power consumption traces of a printer.

It can be observed in the figures that the computer set, electric fan, mobile phone, and laptop have bigger steady state current to peak current (I_{PEAK}) ratios whereas the printer has a very small steady state to peak current ratio, which means that the printer draws a much bigger current upon start-up compared to its current draw during steady state.

The RMS current (I_{RMS}) of an appliance was computed using the average peak current values, per 100 samples, at steady state. This value is central to the computation of an appliance's power consumption, with their relation linear. At steady state, a consistent current consumption pattern for each appliance class can be observed.

A sample power signature of a computer set at steady state is shown in figure 9, with an RMS current value of 0.37881478 A.

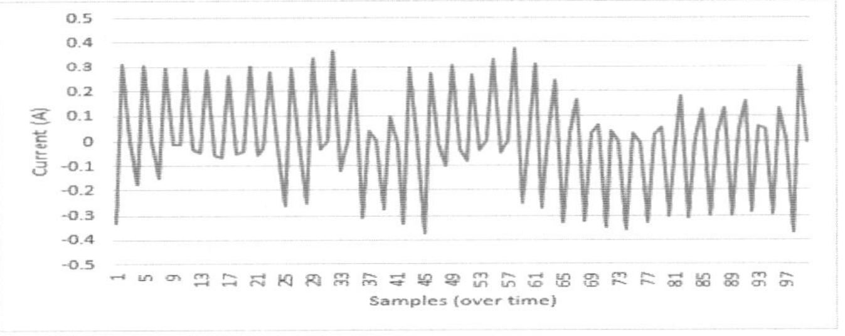

Figure 9. A sample power signature of a computer set at steady state.

A sample RMS current of a mobile phone is 0.074185947 A, evident in its power signature shown in figure 10.

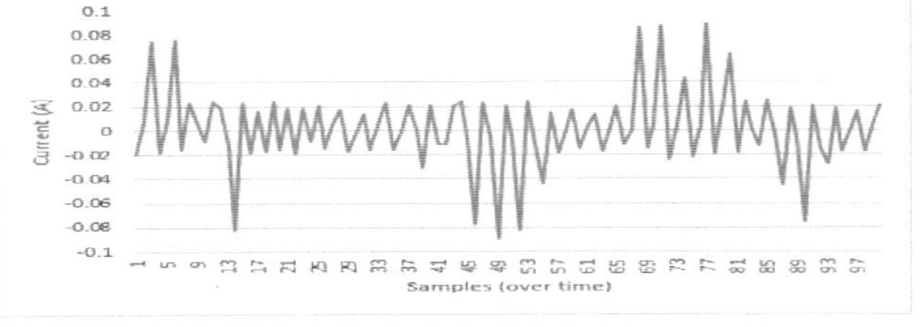

Figure 10. A sample power signature of a mobile phone at steady state.

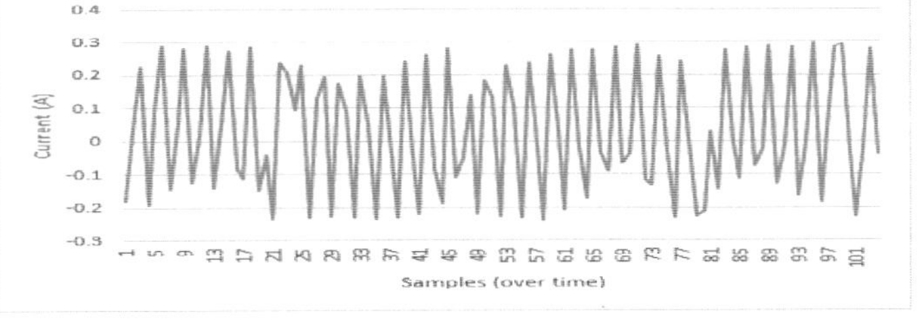

Figure 11. A sample power signature of an electric fan at steady state.

Figure 11 shows a sample power signature of an electric fan at steady state with an RMS current value of 0.276010293 A.

A laptop has a sample RMS current value of 0.339869267 A with its power signature at steady state shown in figure 12.

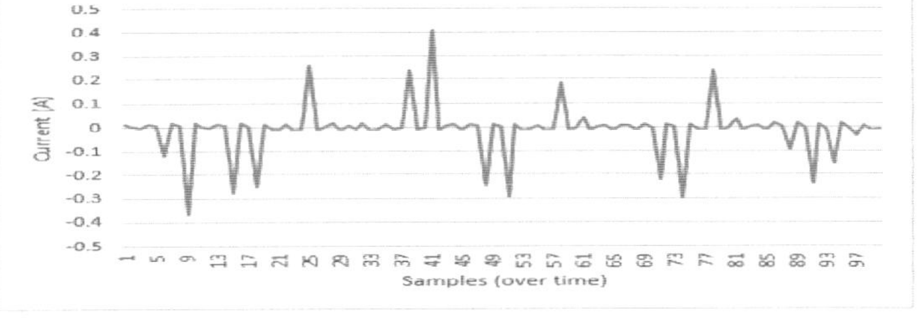

Figure 12. A sample power signature of a laptop at steady state.

A sample power signature of a printer is shown in figure 13, with an RMS current value of 0.127470846 A, a very remarkable drop from 3.704620815 A during start-up.

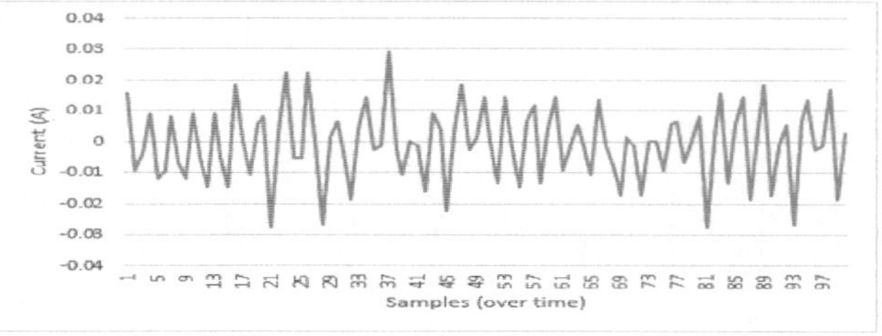

Figure 13. A sample power signature of a printer at steady state.

The value of the real power (P), which is the actual power consumed by an appliance, can only be obtained if the power factor value is known. For the power factor (PF) to be determined, the time difference of t_2 and t_1 in figure 14, that is when the voltage and current signals cross zero, was measured. To increase the measurement accuracy, the average of the zero-crossing time differences for the entire sampling duration was used.

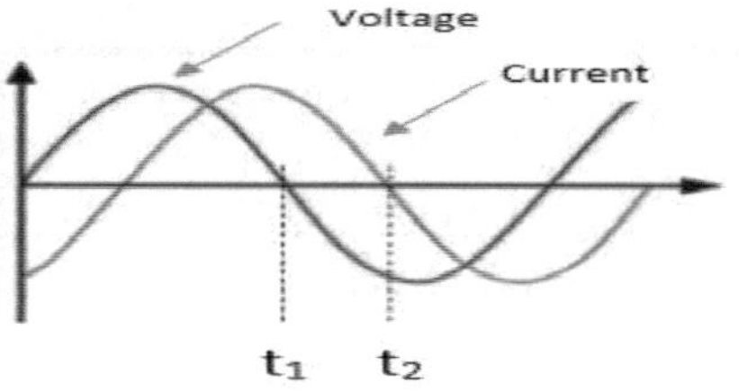

Figure 14. Zero crossings of voltage and current signals.

After all the needed power features were collected and the training set prepared, the k-Nearest Neighbors classifier was tested on the data set with k = 1. The estimated error rate of the classifier on the training set was 0%.

Actual testing results are shown in table 1. The overall classification accuracy is 95.55%. The kNN classifier was able to correctly predict, in all instances, three appliances. The case when the printer was incorrectly classified was when the classification process was initiated and the printer was not turned on immediately, which significantly affected the results of the extracted features from the testing data for classification. The classification model for unknown appliance however slightly affected the overall accuracy with some unknown appliances' power consumption traces closely resembled those of the known classes.

Table 1. Confusion matrix.

	Computer Set	Electric Fan	Laptop	Mobile Phone	Printer	Unknown
Computer Set	14					1
Electric Fan		15				
Laptop			15			
Mobile Phone				15		
Printer				1	14	
Unknown		1		1		13

4. Conclusion and future works

An appliance classification smart meter was developed and a library of five appliance classes, each with fifteen training sets, was created. A supervised learning approach using the kNN classifier was used in the appliance classification, with high accuracy. Based on the results, the timing of data acquisition is critical, especially for appliance whose steady state RMS current to peak current ratio is very small. For future works, the load library needs to be populated with more appliance classes, of different models and brands, to gradually capture all appliance types. Also, for appliance with finite states, the changes in power consumption with state transitions should be considered.

5. References

[1] Reddy, R S, Keesara, N, Pudi, V and Garg, V 2015 *Proceedings of BS2015: 14th Conference of International Building Performance Simulation Association* (India: Hyderabad)

[2] Hart, G W 1992 *Proceeding of the IEEE* 80 12

[3] Ito, M, Uda, R, Ichimura, S, Tago, K, Hoshi, T and Matsushita, Y 2004 *Proceedings of the 2004 IEEE International Symposium on Applications and the Internet* pp 291

[4] Reinhardt, A, Burkhardt, D, Zaheer, M and Steinmetz, R 2012 *Proceedings of the 7th IEEE International Workshop on Practical Issues in Building Sensor Network Applications* (USA: Florida)

[5] Englert, F, Kobler, S, Reinhardt, A and Steinmetz, R 2013 *Proceedings of the 4th International Conference on Future Energy Systems* (USA: California)

[6] Ridi, A, Gisler, C and Hennebert, J 2015 *proceedings of the 13th International Conference on Pervasive Computing and Communications (Percom 2015), Workshop on Pervasive Energy Services (PerEnergy)* pp 323 – 328

[7] Ri di, A, Gisler, C and Hennebert, J 2014 *Proceedings of the 22nd IEEE International Conference on Pattern Recognition* pp 3702 – 3707

[8] Zaidi Adeel Abbas, F K and Palensky, P 2010 *Proceedings of the 36th IEEE Annual Conference on Industrial Electronics Society (IECON)* pp 2442-2447

[9] Kato, T, Cho, H, Lee, D, Toyomura, T and Yamazaki, T 2009 *Ambient Assistive Health and Wellness Management in the Heart of the City* 5597 pp 150–157

[10] Fitta, M, Biza, S, Lehtonen, M, Nieminen, T and Jacucci, G 2010 *Proceedings of the 4th International Conference on Mobile Ubiquitous Computing, Systems, Services and Technologies (UbiComm)* pp 471–477

[11] Saitoh, T, Aota, Y, Osaki, T, Konishi, R and Sugahara, K 2008 *Proceedings of the 23rd International Technical Conference on Circuits/Systems, Computers and Communications (ITC-CSCC)* pp 349–352

[12] Pandiri, S and Srikanth, G 2016 *International Journal of Innovative Technology and Research (IJITR)* 4 4 pp 3350 – 3353

[13] Kurde, A, Kulkarni, V S 2016 *International Journal of Scientific and Research Publications* 6 9

[14] Manjusha, P, Saran Kumar, T, Venkateshwarullu, V and Nageshwar Rao, I R S 2017 *International Journal of Science, Engineering and Technology Research (IJSETR)* 6 9

[15] Murali, M J 2017 *International Journal of Advance Research and Innovative Ideas in Education* 3 3

[16] Cabaccan, C N, Cruz, F R G and Agulto, I C 2017 *Proceedings of the 2017 IEEE 9th International Conference on Humanoid, Nanotechnology, Information Technology, Communication and Control, Environment and Management (HNICEM)* (Philippines: Manila)

[17] Harrington, P 2012 *Machine learning in action* Manning Publications Co. New York

6. Acknowledgements

The researchers wish to extend their sincerest gratitude to all the persons who, in one way or another, contributed to the success of this endeavor. This research was supported by the Commission on Higher Education.

Semantic Mapping and Object Detection for Indoor Mobile Robots

S. Kowalewski, A.L. Maurin, J. C. Andersen
Department of Electrical Engineering, Danmarks Tekniske Universitet, Kgs. Lyngby, Denmark

Abstract. In this paper the authors present a full solution for object-level semantic perception of the environment by indoor mobile robots. The proposed solution not only provides means for semantic mapping but also division of the environment into clusters representing singular object instances. The robot is provided with information that not only allows it to avoid collisions with obstacles present in the environment, but also information about the localization, the class and the shape of each encountered object instance. This level of perception enhances the robot's ability to interact with the environment. The state-of-the-art deep learning solution, Mask-RCNN, is used for the image segmentation task. The image processing network is combined with an RTAB-Map SLAM algorithm to generate semantic pointclouds of the environment. The final part of the paper is focused on pointcloud processing: providing methods for instance extraction and instance processing. To verify the performance of the proposed methodology multiple experiments are conducted. Through the evaluation of the results it is possible to identify possible improvements.

1. Introduction

The environmental awareness is a crucial feature of an autonomously operating robot. The information about surroundings is acquired by available sensors and then processed internally to make autonomous operation in an unknown environment possible. The most basic approach is to use sensors to detect occupied areas and mark them as obstacles. In such case, the robot is aware of the placement of obstacles but lacks any information about the type of the detected objects. This approach is sufficient when the objective is to maneuver and avoid collisions but doesn't provide any additional information, that might be useful when interacting with encountered objects.

The methodology proposed in the paper expands this basic approach by adding an obstacle classification step and instance detection step. In the result, not only the occupancy information is acquired but also the semantic classification for each detected object is determined. Such approach allows to not only navigate around objects but also interact with them depending on the obstacle class; for instance a chair might be moved, whereas a table should be maneuvered around.

The main contributions of this paper are as follows:

- Design of an end to-end solution that provides an object-level classification and outputs class, position and shape of each of the detected objects.
- Verification of the performance through experiments.

The rest of the paper is divided in following manner: in Section 2 related work is reviewed. Section 3 provides the reader with a general overview of the proposed solution and the division of the problem into segments. In sections 4,5 and 6 the methodology used in each of the segments is described. Section 7 includes the description of conducted experiments. The results of experiments are presented in Section 8. Sections 9 and 10 include conclusion and future work, respectively.

2. State of the art

There's been multiple attempts to provide solutions capable of generating 3D semantic maps - maps in which each area is semantically classified. The most common methodology combines a semantic image segmentation solution with a SLAM algorithm.

An example of this approach is SemanticFusion [1] which is a combination of, at the time, the state-of-the-art Deconvolutional Semantic Segmentation network [2] with an ElasticFusion SLAM [3] algorithm and a Bayesian update scheme. This approach allows authors to generate accurate semantic 3D maps of the environment. Nakajima et al. [4] take a similar approach but they assign class probabilities to regions instead of single voxels of the 3D map. This provides a significantly faster solution that can perform at 30Hz frequency. Jeong et al. [5] present a methodology based on the same approach but use multiple sensors to allow outdoor, large scale operation.

A slightly different approach to the segmentation problem is presented by the authors of PointNet [6]. PointNet is a deep neural network that provides end-to-end pointcloud processing capabilities, outputting a variety of results including: object classification, part segmentation and semantic segmentation. Ability to directly process pointclouds is believed to provide superior results. The presented solution is used for single frame processing and authors do not explore the idea of generating global maps.

Nonetheless, these works are focused on the generation of the semantic map and do not attempt the object-level instance division as proposed in this paper.

3. Overview

The proposed system was divided into four essential submodules:

Image acquisition - this module is responsible for the acquisition of RGB and depth images from the Kinect camera.

Image segmentation - this module is responsible for processing acquired images. The performance of this module has a great impact on the whole system as this module adds the semantic meaning to the acquired images. The output images include annotations about detected objects, their classes and localization in the image. Mask-RCNN [7] is used to achieve the module's functionality. For the purpose of the neural network training and performing experiments a GeForce GTX 1080 graphical unit was used.

Robot localization and mapping module - this module is responsible for keeping track of the robot position and simultaneously creating a semantic pointcloud of the environment. RTAB-Map [8] is used to achieve the module's functionality.

Instance detection and processing module - this module is responsible for splitting the created semantic pointcloud into separate object instances. The output of this module is the localization (in the global coordinate system), class and the shape (represented by a pointcloud) of each of detected objects. Moreover, this module is responsible for the instance processing task.

4. Image segmentation

To achieve the semantic image segmentation the Mask-RCNN deep neural network is used. The Mask-RCNN is a neural network architecture created in Facebook AI Research, it expands the capabilities of it's predecessor: Faster-RCNN [9], by adding an image segmentation output. The network processes the input image and provides localization of the classified objects together with segmentation masks.

The original RGB version of Mask-RCNN is modified to process RGB-D images. The reasoning behind the modification is that in many cases the objects seem to be hard to distinguish using only RGB images (for instance when colors of the object and background are similar). In these cases the depth image might provide the missing information necessary to detect object's boundaries and therefore enhance the performance of the network.

The neural network was trained using the SUN-RGBD dataset [10]. This dataset is a compilation of multiple different indoor datasets [11-13] and consists of 10.335 images. It contains 37 different object classes which represent objects that are expected to be found in an indoor environment. The dataset provides comprehensive labeling including all of annotations required for Mask-RCNN training.

The neural network was trained using a sub-set of available classes (chair, table, window, door, box, shelves, sofa, cabinet). The dataset was split into training and validation sets as provided in the dataset information file.

The pre-trained model used for transfer learning was a Mask-RCNN model trained on the the COCO dataset [14]. Multiple training sessions were conducted using RGB and RGB-D variations of Mask-RCNN. For the training sessions in which the Mask-RCNN enhanced with depth was used, the weights of the first convolutional layer of the network could not be transferred as the shape of the convolution was modified. In such cases, this layer was ignored during the transfer learning process.

The results and discussion can be found in Section 8.1.

5. Robot localization and mapping

To be able to map the environment, as the robot is exploring the environment, it is necessary to reliably localize the robot. Moreover, the semantic images generated by the image segmentation module have to be expanded into a pointcloud using the corresponding depth images. These actions are achieved by applying a simultaneous localization and mapping (SLAM) algorithm.

RTAB-Map algorithm was found to be suitable for the task. It is an RGB-D Graph-Based SLAM algorithm that ensures real-time operation. The robot localization and mapping module outputs a semantic 3D pointcloud of the environment together with a 2D costmap of inaccessible areas.

The original implementation of RTAB-Map algorithm had to be adjusted to be suitable for the solution proposed in this paper. A parallel graph was added that includes segmented images instead of the original RGB frames captured during the operation. This approach allows to keep the original operation of the RTAB-Map, which is based on the features extracted from camera frames, and at the same time generate a pointcloud based on semantic images.

6. Instance detection and processing

6.1. Instance detection

The generated semantic pointcloud provides substantial amount of information about the environment but lacks the division into single object instances. To achieve this task, the generated semantic pointcloud of the environment is processed by an euclidean cluster extraction algorithm. The parameters used for the cluster extraction can be found in Tab. 1.

To avoid clustering together points belonging to different classes the full semantic pointclout is divided into subsets corresponding to each of the classes. The euclidean cluster extraction is applied to each of the subsets.

Table 1. Euclidean cluster extraction parameters.

Parameter	Value
Search radius	10[cm]
Minimum size	50[pts]

6.2. Instance processing

The final part of the solution consists of a set of instance processing algorithms. Instance processing is a crucial part of the proposed solution, it illustrates the usefulness of the approach and possible capabilities of a robot with object-level perception of the environment. Two instance processing examples were implemented: chair orientation detection algorithm and approach pose estimation algorithm.

The chair orientation detection algorithm allows the robot to determine the orientation of each of the encountered chair objects. This is beneficial when interacting with chair objects as the robot is aware from which direction the object should be approached.

The approach pose estimation algorithm is used to, given the selected object instance, estimate the target position of the robot when navigating towards the object.

Chair orientation detection algorithm: The algorithm for the detection of the front of the chair is divided into two parts. Firstly, a plane parallel to the z axis is fitted into the instance pointcloud. To find the plane the RANSAC algorithm is used. The detected plane corresponds to the backseat of a

IOP Conf. Series: Materials Science and Engineering **517** (2019) 012012 doi:10.1088/1757-899X/517/1/012012

chair. Obviously this part works only with the assumption that the chair has a distinguishable backseat. The plane provides two normal directions that represent the front and the back of the chair.

Once the backseat plane is detected it is necessary to determine which direction represents the front and which represents the back of the chair. This part of the algorithm works on the assumption that distance-wise, the front of the chair is longer than the back. To determine the orientation of the chair a simple algorithm was proposed: points with maximum distance from the plane are found on the both sides on the plane; the one with a larger distance is determined to be on the front side of the plane.

The detected front direction is then represented with an unit vector placed in the mean point of the instance pointcloud. To allow operations on the 2D obstacle map, the found vector is casted onto the ground plane.

Approach pose estimation algorithm: The approach pose estimation algorithm was introduced to add a capability of navigating the robot towards a selected object instance. To achieve this, the space around the object is sampled by checking in the general obstacle map if the location is accessible. The space is sampled either in ray-like or arc-like fashion. The operation of the algorithm is illustrated in Fig. 1.

7. Benchmarking methods and experiments

7.1. Guidebot

To be able to perform experiments in the real-life scenario and verify the performance of the system it was necessary to use a mobile robot. For this purpose a DTU robot - Guidebot was used. It is a differential drive wheeled mobile robot equipped with a range of sensors including: a Kinect camera (XBOX 360), a laser scanner (SICK LMS200-30106) and wheel encoders.

7.2. Image segmentation

Multiple training sessions were run to provide more insight into the optimal training procedure. To measure the performance of the image segmentation task the mean average precision (mAP@0.5) [15] is used. The results are presented in Section 8.1

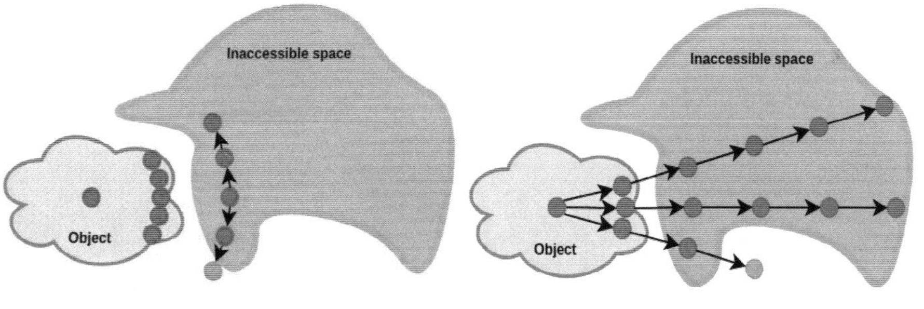

Figure 1. Approach pose estimation algorithm: on the left ray sampling; on the right arc sampling. Red dots represent unsuccessful sampling steps whereas green dots represent the final successful attempt.

7.3. Full solution

A final experiment was proposed to verify the performance of the whole solution in a real life scenario. The experiment consisted of two stages that are described in detail in this section. A single room was setup by placing different recognizable objects at known positions. On Figure 3a it can be seen that to make the task more challenging, a variation of objects was selected for each class (for example round and rectangular tables or different types of chairs). Tab. 2 contains information about objects used in the experiment and their placement. Additionally the last column provides information if the object was detected by the algorithm during the experiment.

During the experiment it was noticed that the robot detects background objects. The localization measurement was not conducted for these objects but for the sake of the completeness of the experiment they are present in the Tab. 2 with ”unknown” coordinates.

During the first stage of the experiment the robot was manually controlled. The goal was to map the robot's environment and create a global semantic pointcloud. Then, the instance detection algorithm provided information about the position and the number of detected objects. These results allow to quantify the performance of the robot by verifying how many of the objects were detected and how precise the solution is.

For the detected objects, a mean localization error was calculated. The error was derived using the following formula, where **d** is the number of the detected objects, x_i^{gt} and y_i^{gt} are the coordinates of the i-th ground truth object, x_i^{pred} and y_i^{pred} are the coordinates of the corresponding predicted object.

$$e = \frac{\sum_{i=1}^{d}[abs(x_i^{gt} - x_i^{pred}) + abs(y_i^{gt} - y_i^{pred})]}{d} \quad (1)$$

Thus, before the start of the second stage of the experiment, the robot has already mapped the environment and is aware of detected objects. During the second stage commands were sent to the robot to approach multiple of the detected objects to see if the robot was able to maneuver in the environment and reach the selected objects.

Table 2. Placement of the objects during the final experiment.

Object	x[m]	y[m]	Detected?
Chair #1	1.25	1	Detected
Chair #2	2.10	1	Detected
Chair #3	3.10	1	Detected
Table #1	2.55	1	Detected
Table #2	0.4	1.2	Detected
Box #1	0.3	1.2	Detected
Box #2	4.45	-0.55	Detected
Box #3	unknown	unknown	Detected
Box #4	unknown	unknown	Detected
Box #5	unknown	unknown	Detected
Shelves #1	3	-0,85	Not detected
Shelves #2	unknown	unknown	Detected
Shelves #3	unknown	unknown	Detected

Table 3. mAP of the best model for each of the classes. It can be observed that classes with more training instances have performed better than classes underrespresented in the dataset.

Class	mAP	# training instances
Chair	0.631	11514
Table	0.53	6913
Window	0.464	2549
Door	0.415	2000
Cabinet	0.399	1464
Sofa	0.334	701
Box	0.229	1982
Shelves	0.203	482

IOP Conf. Series: Materials Science and Engineering **517** (2019) 012012 doi:10.1088/1757-899X/517/1/012012

Table 4. Detected objects and their placement. Green: true positives; Red: false positives; Blue: detection that were present in the environment but not considered during the setup of the experiment

Object	x[m]	y[m]	Notes
Chair #1	1.2	-0.95	Matched
Chair #2	3.18	1.13	Matched
Chair #3	1.94	1.21	Matched
Chair #4	2.57	-0.98	Missclassification
Table #1	2.36	0.94	Matched
Table #2	0.57	1.18	Matched
Table #3	3.17	-1.04	Missclassification
Table #4	5.13	1.05	Missclassification
Table #5	4.98	0.81	Missclassification
Window #1	5	1	Missclassification
Window #2	1.29	-1.46	Missclassification
Box #1	0.33	1.31	Matched
Box #2	4.87	0.4	Detected background object
Box #3	4.27	-0.3	Matched
Box #4	5.22	0.63	Detected background object
Box #5	4.95	1.81	Detected background object
Shelves #1	1.62	-1.47	Detected background object
Shelves #2	4.96	0.35	Detected background object

8. Results

8.1. Segmentation results

To find a model with an acceptable performance, the training sessions were conducted using both: the RGB architecture and the RGB-D architecture. The compilation of performed training sessions can be seen in Fig. 2.

The best performing model achieved the mAP of 0.414 (perfect mAP score is 1). The visual inspection of the results has proven a good performance of the network. The model was trained for 90 epochs (450000 iterations). SGD optimization was used with a learning rate of 0.001. The architecture of the best performing model was designed for RGB images as input. A more detailed analysis has been performed for the best performing model: Tab. 3 shows the performance of the model and number of training examples for each of the classes.

Unexpectedly, the Mask-RCNN with RGB-D inputs has achieved inferior results (maximum mAP 0.386). The reason for this might be the fact that the first convolutional layer was ignored during the transfer learning procedure. Moreover, for each hyperparameters setting only a single training session was conducted. Due to the stochastic nature of neural networks a single training result does not provide a definite answer about performance differences.

8.2. Full solution results

Tab. 4 provides information about objects detected during the experiment.

Firstly, it was observed that the final solution achieved a high recall value - detecting nearly all of the present objects. The recall value for the class, table and box classes reaches 1.0 what means that all of the ground truth objects were detected. For the shelf class the recall value drops to 0.67 due to the fact that one of the shelves was not detected.

The robot performs noticeably worse when the precision value is taken into account. This means that the final solution detects multiple objects that were not present in the environment. The robot detected 1 chair, 3 tables and 2 windows that were not present in the prepared room. It is worth noticing pointclouds for the majority of the miss-classified instances consist of small number of points. This means that the solution might be improved by changing the pointcloud acceptance threshold in the instance detector.

The calculated mean localization error value (see Formula 1) is 0.25[m]. It is expected that even better results could be achieved if the mapping process was extended to allow the robot to see the object from increased number of viewpoints.

For the second stage of the experiment, three detected object instances from the following classes were chosen: a chair, a table, a box. For each of the tested object the solution provided positions and orientations that would result in the robot facing the selected obstacle. The determined pose was visualized on the screen and was correct. Using a Robot Operating System (ROS) navigation package it was possible for the robot to avoid collisions and approach the selected object, given a desired end pose.

9. Conclusion

The main objective of the paper was to provide an semantic mapping and instance detection solution for indoor mobile robots.

The methodology has been designed, implemented and verified. The final experiment has shown that the robot equipped with the proposed system was able to map it's environment and assign semantic meaning to sections of the map. Moreover, the robot was able to divide the environment into single instances of objects belonging to various classes - keeping track of their localization and shape. Additionally, it has been proven that the acquired information about the shape of objects can be reliably used to determine orientation of objects. Finally, it was possible to use the acquired information to autonomously navigate the environment and approach objects.

To improve the results of the system, a detailed analysis of the performance of the image segmentation module has been conducted. A modification to the RGB Mask-RCNN implementation has been introduced by adding a depth channel to the input. No improvement of the performance was observed for the network with RGB-D inputs.

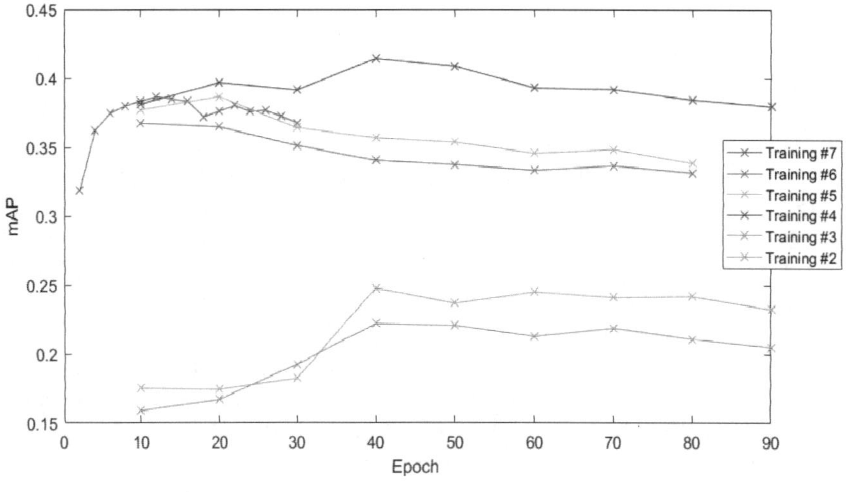

Figure 2. Performed training sessions. During the training sessions #2 and #3 the initial convolutional layers were frozen and therefore the performance of these models is inferior. After noticing that the models tend to overfit during long training procedures, session #7 was ran shorter with a finer performance sampling.

IOP Conf. Series: Materials Science and Engineering **517** (2019) 012012 doi:10.1088/1757-899X/517/1/012012

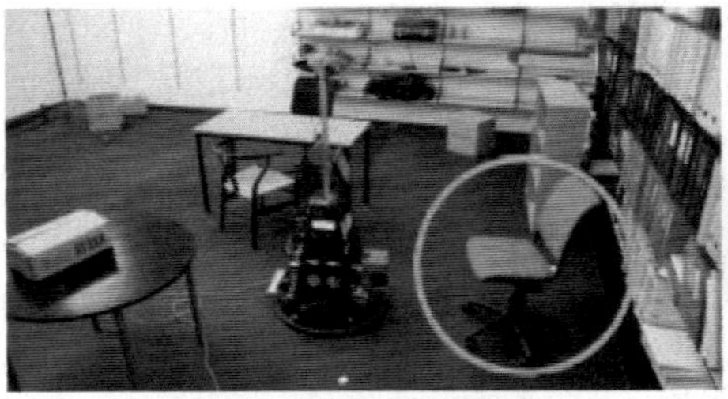

(a) The experiment enviroment

(b) Semantic pointcloud viewpoint A

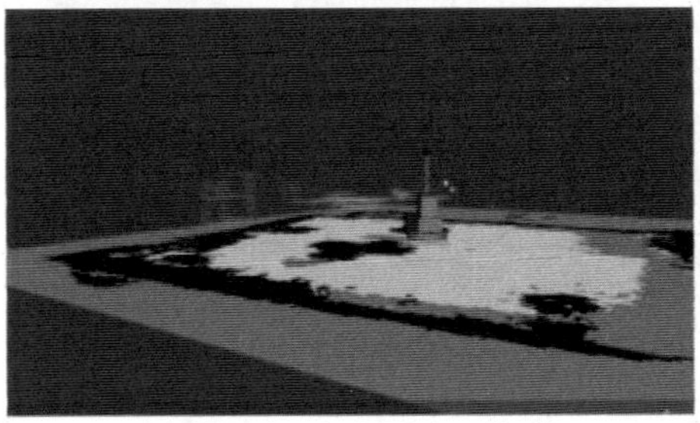

(c) Semantic pointcloud viewpoint B

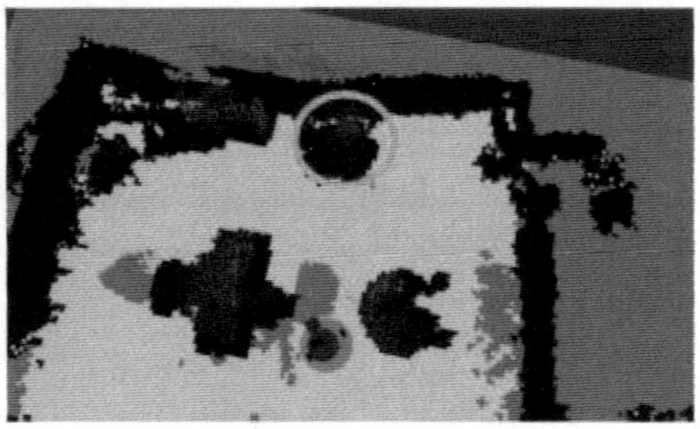

(d) Semantic pointcloud viewpoint C

Figure 3: The experiment environment and the generated semantic pointcloud seen from different viewpoints. Light green - chairclass; Dark green - box class; Red points - table class; Blue - shelves class. To make scene interpretation easier for the reader,one of the chairs is marked with a green circle

10. Future work

Although the results of experiments are satisfying, there are possibilities to further improve the performance of the solution. A more insightful Mask-RCNN training procedure should be conducted to provide a definite answer about the performance of the RGB and RGB-D Mask-RCNNs. Additional training sessions should be run for original and modified versions of the network to remove the impact of performance variance. Moreover, the performance of a neural network depends strongly on the number of available training examples. Throughout experiments it was proven that many of the classes available in the used dataset were underrepresented and hence the performance for these classes was inferior. A different dataset might be used or the existing dataset might be expanded to improve the quality of the image segmentation process.

The used SLAM algorithm (RTAB-Map) adds a significant limitation to the solution - the environment has to be static. In the future, it would be reasonable to replace the current SLAM algorithm with one that is suitable for operation in a dynamic environment. This would greatly expand capabilities of the solution.

Only a few scenarios for instance processing were presented. It is believed, that the paper provides a foundation to implement various instance processing algorithms - allowing the robot to approach and interact with different classes of objects. In the future, a set of behaviours should be implemented for each of the detectable object classes.

11. References

[1] Janoch, A., Karayev, S., Jia, Y., Barron, J. T., Fritz, M., Saenko, K. and Darrell, T. 2011. "*A category-level 3-D object dataset: Putting the Kinect to work*" in *2011 IEEE International Conference on Computer Vision Workshops (ICCV Workshops).*

[2] Qi, C. R, Su, H., Mo, K.and Guibas, L. J. 2016. "*PointNet: Deep Learning on Point Sets for 3D Classification and Segmentation*" *CoRR,* vol. abs/1612.00593.

[3] Noh, H., Hong, S. and Han, B. 2015. "*Learning Deconvolution Network for Semantic Segmentation*" *CoRR,* vol. abs/1505.04366.

[4] Jeong, J., Yoon, T. and Park, J. B. 2018. "*Multimodal Sensor-Based Semantic 3D Mapping for a Large-Scale Environment*" *CoRR,* vol. abs/1802.10271.

[5] McCormac, J., Handa, A., Davison, A. J. and Leutenegger, S. 2016. "*SemanticFusion: Dense 3D Semantic Mapping with Convolutional Neural*" *CoRR,* vol. abs/1609.05130.

[6] Xiao, J., Owens, A. and Torralba, A. 2013. "*SUN3D: A Database of Big Spaces Reconstructed Using SfM and Object Labels*" in *2013 IEEE International Conference on Computer Vision*

[7] He, K., Gkioxari, G., Dollár, P. and Girshick, R. B. 2017. "*Mask R-CNN*" *CoRR,* vol. abs/1703.06870.

[8] Everingham, M., Gool, L., Williams, C. K., Winn, J. and Zisserman, A. 2010. "*The Pascal Visual Object Classes (VOC) Challenge*" *Int. J. Comput. Vision,* vol. 88, no. 2, pp. 303--338.

[9] Labbé, M. "*Real-Time Appearance-Based Mapping*" 2018. [Online]. Available: http://introlab.github.io/rtabmap/. [Accessed 21 06 2018].

[10] Matterport, "*Mask-RCNN model pre-trained on COCO dataset*" [Online]. Available: https://github.com/matterport/Mask_RCNN/releases/download/v2.0/mask_rcnn_coco.h5. [Accessed 21 06 2018].

[11] Silberman, N., Hoiem, D., Kohli, P. and Fergus, R 2012. "*Indoor Segmentation and Support Inference from RGBD Images*" in *ECCV.*

[12] Ren, S., He, K., Girshick, R. B. and Sun, J., 2015. "*Faster R-CNN Towards Real-Time Object Detection with Region Proposal Networks*" *CoRR,* vol. abs/1506.01497.

[13] Song, S., Lichtenberg, S. P. and Xiao, J. 2015. "*SUN RGB-D: A RGB-D Scene Understanding Benchmark Suite*" in *The IEEE Conference on Computer Vision and Pattern Recognition (CVPR).*

[14] Whelan, T., Salas-Moreno, R. F., Glocker, B., Davison, A. J. and Leutenegger, S. 2016. "*ElasticFusion: Real-time dense SLAM and light source estimation*" *The International Journal of Robotics Research,* vol. 35, no. 14, pp. 1697-1716.

[15] Nakajima, Y., Tateno, K., Tombari, F. and Saito, H. 2018. "*Fast and Accurate Semantic Mapping through Geometric-based Incremental Segmentation*" *CoRR,* vol. abs/1803.02784.

IOP Conf. Series: Materials Science and Engineering **517** (2019) 012013 doi:10.1088/1757-899X/517/1/012013

The Gapless Design for High Sensitivity of Current Mode Dual Magnetodiode

Chalin Sutthinet[1], Sawatdipong Poonsawat[1], Toempong Phetchakul[1] and Amporn Poyai[2]

[1] Electronics Department, Faculty of Engineering, King Mongkut's Institute of Technology Ladkrabang, Bangkok, Thailand
[2] Thai Microelectronics Center, National Electronics and Computer Technology Center, Chachoengsao, Thailand

Abstract. This paper presents the design for high sensitivity of magnetic detector device in current mode that uses Lorentz's force deflects current depend on the magnetic density via the dual magnetodiode structure. The current mode devices have two symmetry regions for carrier current receiving that injected from another opposite region. This design has a gap between two carrier current receiving regions that causes some loss and reduce sensitivity. The proposed design has one carrier current receiving region or gapless design that has no loss from gap. The sensitivities of gap 5, 2.5 and 0 μm at 1 mA are 0.0010, 0.0028 and 0.0065 T^{-1} for split cathode structure and 0.00084, 0.0020 and 0.0051 T^{-1} for split anode structure, respectively. This design can be applied to all current mode magnetic device for high sensitivity.

1. Introduction

The magnetic sensors are the devices for detect magnetic field. There are various types and mechanisms [1], [2], [3], [4]. The Hall effect based devices are a group that detect magnetic field by this effect by ferromagnetic and non-ferromagnetic material (metal and semiconductor) [5]. There are two modes of operation that are voltage mode and current mode. The voltage mode use Hall voltage that is induced from magnetic field cross with current for balancing force between electric force qE_H and Lorentz's force $F_L = qv \times B$ where q is electric charge, E_H is Hall electric field, v is drift velocity of carrier and B is magnetic field density. The current mode use the Lorentz's force F_L that greater than electrical force from Hall voltage deflect current or carrier. The structure of this mode device must have two separate regions for receiving the differential current which drain from another region. The output of current mode device is differential current ΔI of the two regions. Magnetotransistors are the transistor that detects the magnetic field [6]. They are both in voltage and current mode of Hall effect. In current mode device, the collector must design in split collectors for receiving injected current passed base from emitter. The output is the differential collector current ΔI_c which will have some value depend on the magnetic field density and direction.

The MAGFET (Magnetic Field Effect Transistor) [7] is one of the current mode Hall effect magnetic field detectors. The structure is split drain for receiving current from source. The output is the differential drain current ΔI_D. This device shows output $\Delta I_D = 0$ when no applied magnetic field. When magnetic field is applied, $\Delta I_D \neq 0$ depend on the density and direction.

Dual magnetodiode is the diode that detect magnetic field [8]. The structure has one anode and two separate cathodes. The current injected from anode to cathodes. The output is the differential current of cathodes ΔI_K. When the magnetic field is applied, the output $\Delta I_K \neq 0$ and when no magnetic field is applied $\Delta I_K = 0$.

From the example that mention above, they use the Lorentz's force deflect current from source of current to receiving regions, collectors, drains and cathodes. The structure of each device is designed for split regions that should have a gap between them. Even though, they are designed in very small scale technology as possible but the gap still exist in these structures. This design is used for very long time since the first device was created [9]. This paper will present the smarter design for higher sensitivity of these devices. The proposed design has no gap or gapless and independent of fabrication technology scaling

2. Device structures and TCAD

The structures of dual magnetodiode studied here are shown in Fig. 1. The name dual diode [8] is from its structure that looks like two couple diodes that have one common anode and split cathode as shown in Fig. 1(a) or one common cathode and split anode as shown in Fig. 1(b).

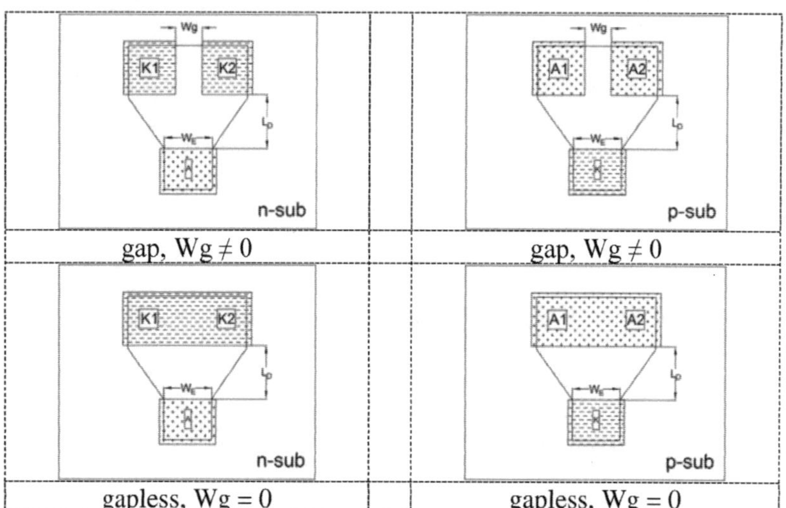

Figure 1. The structures of dual magnetodiode.

The reason why it has to split cathode or anode is for magnetic field detection in current mode. The split cathodes one is start from n-type substrate, 10^{15} cm^{-3}, and diffuse acceptor atoms for p+ anode, 10^{20} cm^{-3}, on substrate and split anode one is start from p-type substrate, 10^{15} cm^{-3} and diffuse donor atoms for n^+ cathode, $5x10^{19}$ cm^{-3}, on substrate. The ohmic contacts of substrate are formed by diffusion the same dopant atom as substrate in very high concentration. Each anodes and cathodes has a contact in the middle of region. The gap Wg is the distance between split regions, the distance between anode and cathode is L_D and the width of injection carrier cross section area of diode is W_E which all of them are shown in Fig. 1. In this study, the W_E is fixed constant at 8 μm, L_D is fixed at 10 μm but the gap Wg are varied in three values of 5, 2.5 and 0 μm. In the case of Wg= 0 μm is called gapless design which are also shown in Fig. 1.

TCAD sentaurus [10], [11], [12] is one of standard commercial programs for device and process simulation in micro/nano electronic industries.

It has magnetic models for magnetic response of device. The current density with magnetic field dependence model for this study is explain as

$$\vec{J}_\alpha = \mu_\alpha \vec{g}_\alpha + \mu_\alpha \frac{1}{1+\left(\mu_\alpha^* B\right)^2}\left[\mu_\alpha^* \vec{B}\times \vec{g}_\alpha + \mu_\alpha^* \vec{B}\times\left(\mu_\alpha^* \vec{B}\times \vec{g}_\alpha\right)\right] \tag{1}$$

where α is n or p type of semiconductor material, $\vec{J}_\alpha$ is carrier current density, $\vec{g}_\alpha$ is current vector without mobility, μ_α is drift mobility, μ_α^* is the Hall mobility, $\vec{B}$ is the magnetic induction vector and B is the magnitude of vector. This magnetic model with the conventional model of semiconductor material is used for dual magnetodiode in this work for studying the device design and mechanism for high sensitivity [13], [14].

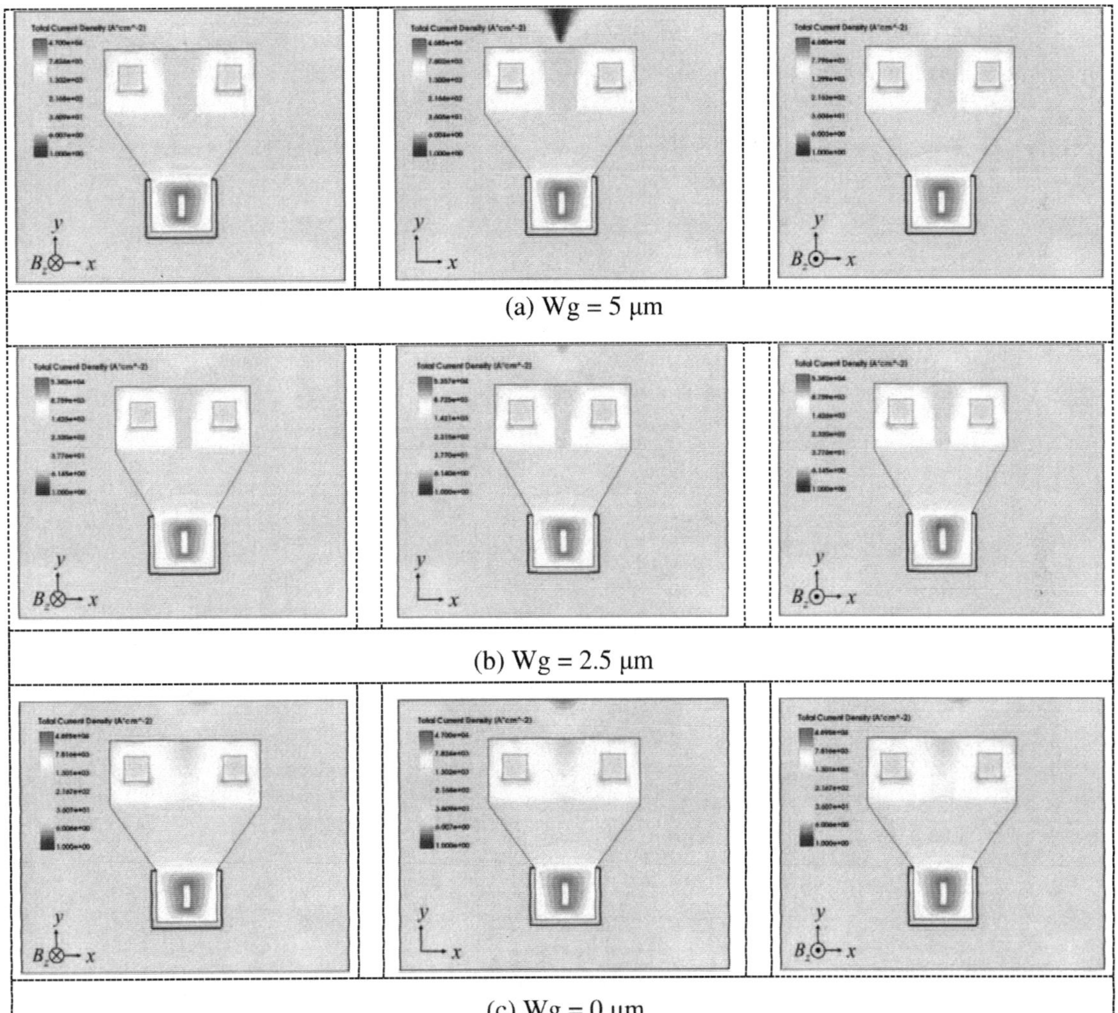

(a) Wg = 5 μm

(b) Wg = 2.5 μm

(c) Wg = 0 μm

Figure 2. Current density distribution of split cathode Dual Magnetodiode.

3. Experiment and Results

The magnetodiode split cathode and split anode structures as shown in Fig.1 are studied by TCAD simulation [15], [16]. These diodes are biased in forward direction by constant current source for keeping amount of carriers in order to avoid carrier modulation from magnetic field. The current density distribution are extracted out and shown as results. Then the magnetic field is applied in $-z$ and $+z$ direction and they show the variation of current density distribution with magnetic field. Dual magnetodiode is the current mode device that the output is differential diode current ΔI_D (differential cathode current for split cathode diode or differential anode current for split anode diode) which varied with magnetic field in amplitude and direction. Figure 2 shows current density distribution of split cathode dual magnetodiode with and without magnetic field. The results of cathode gap Wg = 5, 2.5 and 0 μm are shown in Fig. 2 (a), (b, (c), respectively. The magnetic field in $-z$ direction is applied; the diode currents tilt to the left cathode.

On the other hand, when the magnetic field in $+z$ direction is applied; the diode currents tilt to the right. The cathodes current are equal when no applied magnetic field. Figure 4 (a) is the plot of ΔI_D versus magnetic field B of split cathode diode. ΔI_D is I_1 - I_2 where I_1 is left cathode current and I_2 is right cathode current so the slope of graphs are positive. The relative sensitivities $S_R = \frac{1}{I_D} \cdot \frac{\Delta I_D}{\Delta B}$ at 1 mA of gaps 5, 2.5 and 0 μm are 0.0010, 0.0028 and 0.0065 T^{-1}, respectively.

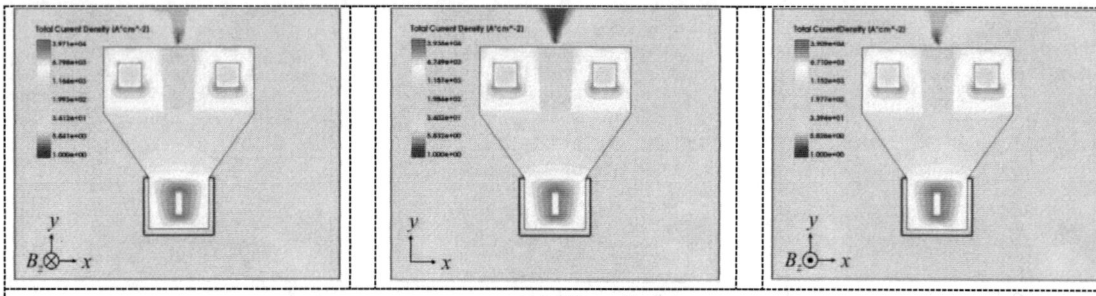

(a) Wg = 5 µm

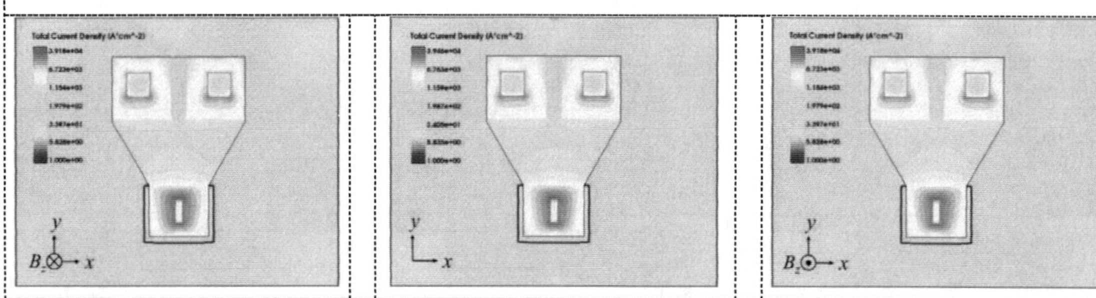

(b) Wg = 2.5 µm

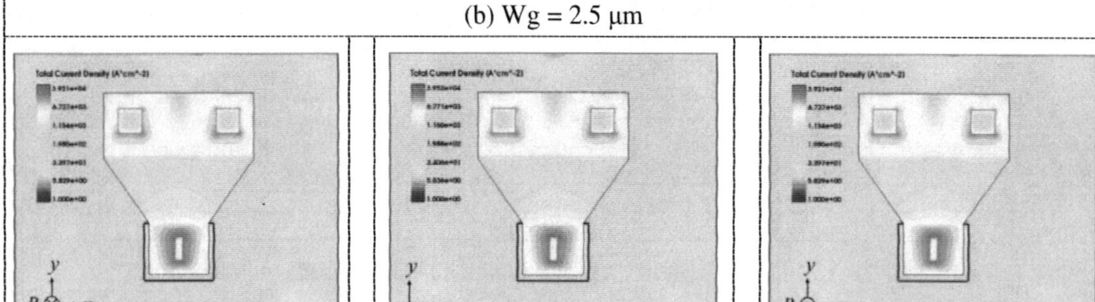

(c) Wg = 0 µm

Figure 3. Current density distribution of split anode Dual Magnetodiode.

Figure 3 is the current density distribution of split anode diode structure. The same condition of forward biased, magnetic field and gaps parameters are applied. The conventional currents flow from split anodes to cathode. The plot of ΔI_D versus magnetic field B of split anode is shown in Fig. 4(b). In this case ΔI_D is I_1 - I_2 where I_1 is left anode current and I_2 is right anode current. The positive slopes means that when magnetic field $+z$ direction is applied, the anode current I_1 is greater than I_2 and when magnetic field $-z$ is applied, the anode current I_2 is greater than I_1. The relative sensitivities $S_R = \frac{1}{I_D}.\frac{\Delta I_D}{\Delta B}$ at 1 mA of gaps 5, 2.5 and 0 μm are 0.00084, 0.0020 and 0.0051 T^{-1}, respectively.

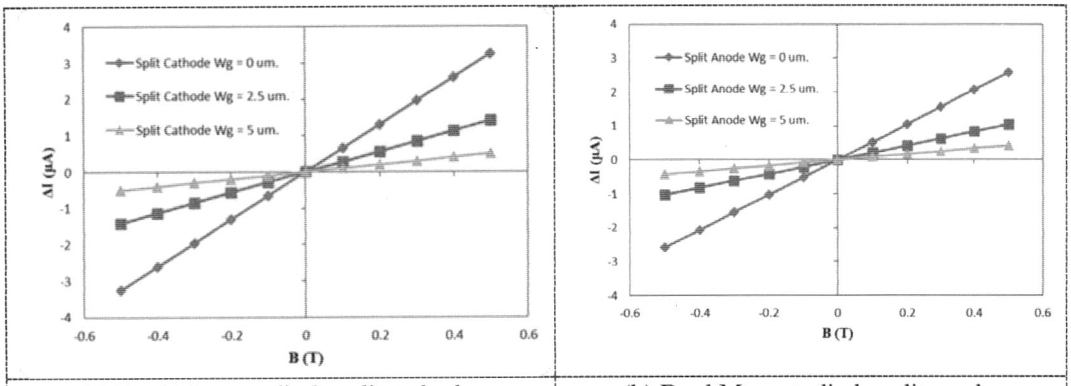

(a) Dual Magnetodiode split cathode. (b) Dual Magnetodiode split anode.

Figure 4. The magnetic field response of Dual Magnetodiode.

4. Discussion

Dual magnetodiode structure studied here is the normal diode that has one common anode and split cathode or common cathode and split anode. This special design is used for magnetic detection field rather than use as rectify. The current of dual magnetodiode is the same as p-n junction diode that can be written as

$$J = \left[\frac{qD_p p_{n0}}{L_p} + \frac{qD_n n_{p0}}{L_n}\right]\left[\exp\left(\frac{qV_a}{kT} - 1\right)\right] \quad (2)$$

where q is electronic charge, D_n and D_p are minority carrier electron and minority carrier hole diffusion coefficient, L_n and L_p are minority carrier electron and hole diffusion length, pn0 and np0 are thermal equilibrium minority carrier electron and minority carrier hole concentration, k is Boltzmann's constant, T is temperature and V_a applied voltage across junction. Equation 2 shows the amount of total current that calculate from the minority carrier diffusion current densities at the edge of space charge region. In the case of p^+n or split cathode n-substrate, the term $n_{p0} << p_{no}$ and can be neglected. The minority diffusion current is approximately hole injected across junction and become minority carrier which then diffuse from junction to split cathode terminals and recombine with majority carrier. The minority carrier diffusion current decay exponentially with distance for several time of minority diffusion length L_p before completely recombine. In order to supply the majority electrons that are lost by recombination with the injected excess minority carrier hole and supply electrons that are being injected across the junction into p region, the electron drift into cathode create the majority electron drift current. The same discussion applies to the drift of holes majority carrier in p region in the case of pn^+ or split anode p-substrate [17], [18].

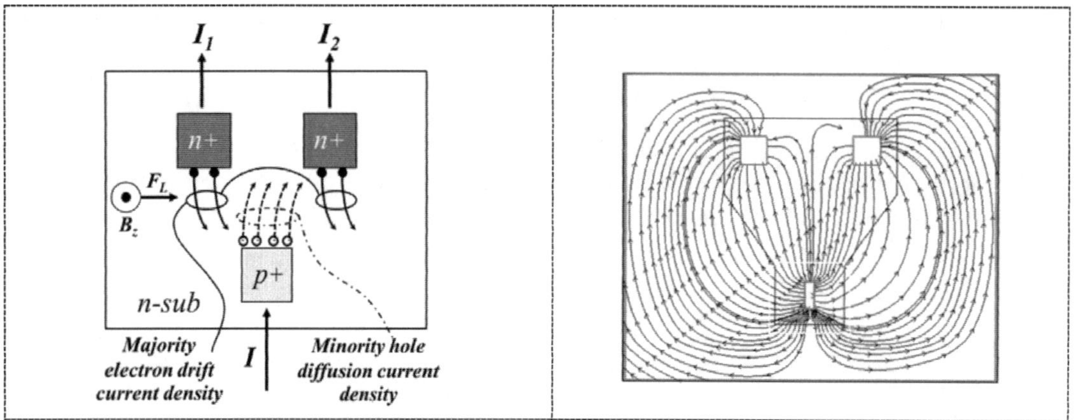

(a) Model of hole minority and electron majority carrier current density deflection.

(b) Conventional total current density deflection vector.

Figure 5. Current deflection of split cathode when applied magnetic field in +z direction.

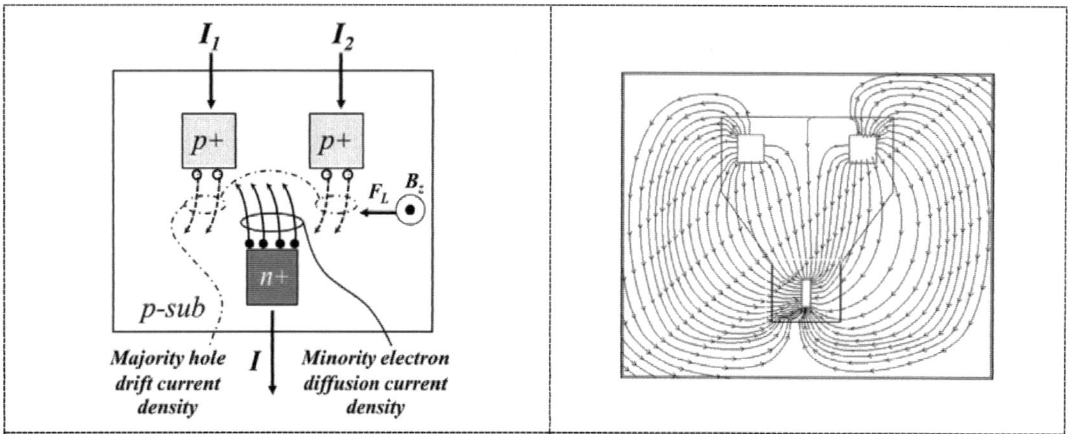

(a) Model of electron minority and hole.

(b) Conventional total current density deflection vector.

Figure 6. Current deflection of split anode when applied magnetic field in +z direction.

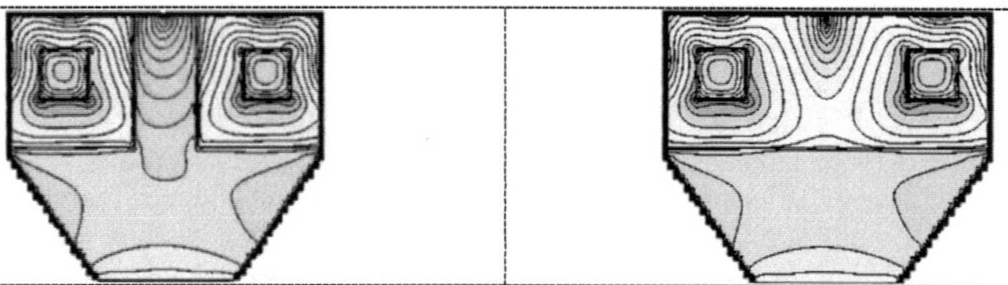

(a) Gap Wg = 5 μm.

(b) Gapless Wg = 0 μm.

Figure 7. Comparison of total current density distribution of split cathode.

Figure 5 shows current deflection in split cathode dual magnetodiode when magnetic field is applied in $+z$ direction. As mention before, there are two carrier current components. The holes are injected from p^+ anode across junction to cathode and diffuse in n-substrate to cathodes. The Lorentz's force is induced in x direction causes the minority hole diffusion current density J_h tilts in the right cathode. The hole minority diffusion current density J_h of right cathode is greater than the left one. The electrons drift into cathode to supply the electron recombine with excess holes create majority electron drift current density J_e from cathode to junction. The electron current density J_e tilts in x direction because of Lorentz's force. The majority electron current density J_e of left cathode is greater than the right cathode. They are shown in Fig. 5(a). Figure 5(b) is the conventional total current density vector of carrier currents in Fig.5 (a) by TCAD simulation. The total currents are the sum of electron and hole currents. As the results $\Delta I = I_1 - I_2$ shown in Fig.4, I_1 is greater than I_2 in this case.

Figure 6 shows current deflection of split anode when applied magnetic field in $+z$ direction. In similarly, the model of carrier deflection and conventional total current density deflection vector are shown in Fig. 6 (a) and (b), respectively.

Figure 7 shows the total current density distribution selected from Fig.2. They are split cathode diodes with gap Wg = 5 μm and gapless Wg = 0 μm with the magnetic field $+z$ direction. The deflection current density with gap in Fig.7 (a) is not smooth and continuous. It is observed that there is low current density area in gap. The high current density deflection does not tilt continuous between cathodes. There is some loss of output differential current. The gapless in Fig. 7 (b) shows the smooth and continuous current density during deflection. There is no low current density area inserted in high current density area. The current density tilts smoothly and continuously. There is no loss and approach to the ideal. In split anode and magnetic field in $-z$ direction, it can be explained similarly.

5. Summary

Dual magnetodiode is a current mode device that can detect magnetic field. The dual magnetodiodes for this study are split cathode and split anode structure. It is current composed of two current components. One is minority diffusion current from injected holes from p^+ anode for split cathode and from injected electron from n^+ cathode for split anode across the junction, respectively. Another is majority electron drift current in cathode n-substrate for split cathode or majority hole drift current in anode p-substrate for split cathode. The induced Lorentz's force acts upon both currents. The current mode device use Hall effect in current mode for current deflection. These devices have to be designed with split current receiving region to support current deflection from magnetic field. It must have a gap between split regions and has some loss in differential current ΔI. The gapless design is proposed here for high sensitivity. It is not necessary to split receiving carrier current region anymore because the current deflection effect will be continue within this region until reach to the inner contacts. From this study, the sensitivity of gapless design is highest both in two structure. The relative sensitivity of gap 5, 2.5 μm and gapless are 0.0010, 0.0028 and 0.0065 T^{-1} for split cathode 0.00084, 0.0020 and 0.0051 T^{-1} for split anode, respectively. The gapless design is the ideal for the high sensitivity current mode device.

6. References

[1] R. S. Popovic, H. P. Baltes, and F. Rudolf, "An integrated silicon magnetic field sensor using the Magnetodiode principle," IEEE Transactions on Electron Devices, vol. 31, pp. 286-291, 1984.
[2] E. H. Putley, The Hall effect and semi-conductor physics, 1968.
[3] E. Ramsden, Hall-effect sensors: theory and application: Elsevier, 2011.
[4] R. S. Popovic, Hall effect devices: magnetic sensors and characterization of semiconductors: CRC Press, 2004.
[5] P. Grunberg, "Magnetic field sensor with ferromagnetic thin layers having magnetically antiparallel polarized components," ed: Google Patents, 1990.
[6] P. Sottip, T. Phetchakul, C. Leepattarapongpan, N. Penpondee, P. Pengpad, A. Srihapat, et al., "The low power magnetotransistor based on the CMOS technology," in ECTI-CON2010: The 2010 ECTI International Confernce on Electrical Engineering/Electronics, Computer, Telecommunications and Information Technology, 2010, pp. 764-767.
[7] E. Yosry, W. Fikry, A. El-henawy, and M. Marzouk, "Compact model of dual-drain MAGFETs simulation," Int. J. Electron. Commun. Comput. Eng, vol. 1, pp. 112-116, 2009.
[8] T. Phetchakul and S. Junkamkaw, "Dual magnetodiode," in Proceedings of the 2009 12th International Symposium on Integrated Circuits, 2009, pp. 502-504.
[9] L. Davies and M. Wells, "Magneto-transistor incorporated in a bipolar IC," in Proc. ICMCST, Sydney, Australia, 1970, pp. 34-35.
[10] Sentaurus Process User Guide. Mountain View, CA, USA: Synopsys Inc, 2016.

[11] Sentaurus Structure Editor User Guide. Mountain View, CA, USA: Synopsys Inc, 2016.
[12] Sentaurus Device User Guide. Mountain View, CA, USA: Synopsys Inc, 2016.
[13] M. Crescentini, M. Biondi, A. Romani, M. Tartagni, and E. Sangiorgi, "Optimum design rules for CMOS Hall sensors," Sensors, vol. 17, p. 765, 2017.
[14] A. Nathan, A. M. J. Huiser, and H. P. Baltes, "Two-dimensional numerical modeling of magnetic-field sensors in CMOS technology," IEEE Transactions on Electron Devices, vol. 32, pp. 1212-1219, 1985.
[15] C. Leepattarapongpan, T. Phetchakul, N. Penpondee, P. Pengpad, A. Srihapat, W. Jeamsaksiri, et al., "The Effect Injection Width and Temperature-Offset Compensation of Magnetotransistor," Procedia Engineering, vol. 25, pp. 1269-1272, 2011/01/01/ 2011.

[16] C. Riccobene, G. Wachutka, J. Burgler, and H. Baltes, "Operating principle of dual collector magnetotransistors studied by two-dimensional simulation," IEEE Transactions on Electron Devices, vol. 41, pp. 1136-1148, 1994.
[17] J.-P. Colinge and C. A. Colinge, Physics of semiconductor devices: Springer Science & Business Media, 2005.
[18] K. Hess, "Advanced theory of semiconductor devices," 2000.

7. Acknowledgments

The authors would like to thank National Research Council of Thailand for financial support of this research.